HUNTSVILLE
TEXTILE MILLS & VILLAGES

HUNTSVILLE
TEXTILE MILLS & VILLAGES

Linthead Legacy

TERRI L. FRENCH

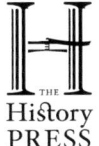

Published by The History Press
Charleston, SC
www.historypress.net

Copyright © 2017 by Terri L. French
All rights reserved

First published 2017

Manufactured in the United States

ISBN 9781467137089

Library of Congress Control Number: 2017931802

Notice: The information in this book is true and complete to the best of our knowledge. It is offered without guarantee on the part of the author or The History Press. The author and The History Press disclaim all liability in connection with the use of this book.

All rights reserved. No part of this book may be reproduced or transmitted in any form whatsoever without prior written permission from the publisher except in the case of brief quotations embodied in critical articles and reviews.

CONTENTS

Forewords, by Mayor Tommy Battle and Marcia Freeland — 7
Preface — 9
Acknowledgements — 11

1. The Southern Migration of New England's Cotton Mills — 13
2. The First Huntsville Cotton Mills — 27
3. Dallas Mill — 36
4. Lincoln Mill — 48
5. Merrimack Mill — 59
6. Lowe Mill — 73
7. The East Huntsville Addition — 82
8. Dangers, Unrest and Upheaval — 84
9. Preservation, Restoration and Revitalization — 96

Chronology — 119
Notes — 123
Bibliography — 131
Index — 139
About the Author — 143

FOREWORDS

Cotton mills transformed Huntsville's agrarian landscape into a thriving industrial city during the early turn of the century. It was because of this industry that Huntsville saw a sharp increase in its prosperity. Huntsville owes much to the innovators of that era, but it owes even more to the everyday men and women who helped to grow and shape this city.

Little remains of this bygone time beyond buildings and memories. It is important for us to acknowledge the importance of this era in history and to preserve the contributions of the early Huntsvillians that help to shape this city. Author Terri French has beautifully reminded us of the rich history emanating from our cotton roots. She puts a face on the people whose names claim the streets and buildings we encounter each day. Huntsville has always been a city of hard work, perseverance and fortitude, and we are grateful to Ms. French for capturing the spirit of our earliest industrial days with this thoughtful record.

—Mayor Tommy Battle

Lowe Mill is such a grand place.

I am grateful to the Hudsons for saving this old building and feel fortunate to work in a historic cotton mill where the paint is peeling and I can still find panes of glass that were handmade. The light from the windows is beautiful; I often walk the floors of the mill in the mornings when there are few people in studios. It is peaceful in the morning.

FOREWORDS

The roof construction is splined. Look it up! I cannot fathom the strength it took to put it together, but I love looking up at the huge beams. The cracks between the floor planks still hold treasures from the past. The planks are still soaked with oil; when the sun hits the floors, little pools form.

Former workers have talked to me about how the windows were all open with fans and machines running so loud they couldn't talk to the person beside them. I feel spoiled having air conditioning and heat. Galleries have replaced the water stations and salt tablets at the ends of the buildings. We now have cold water fountains; again, I feel spoiled. The generation that came before me was tough.

If the artists or patrons complain about the heat or the cold or anything really, I just smile. Partly because of how prosperous the mill is, partly because I imagine throwing them off the water tower.

I am so thankful that there are individuals like Terri French who are investigating the history and culture of the former mills. It is wonderful to work in such a thriving creative space where individuals and art inspire. There is collaboration among the artists and the patrons in this incubator, and I get to help others achieve their dreams.

—Marcia Freeland, executive director
Lowe Mill ARTS & Entertainment

PREFACE

Many folks attribute the old adage "write what you know" to Hemingway. I don't know where or from whom the saying originated, but I do know I don't entirely agree with it. I've always been a curious sort, questioning adults and so-called authorities and, well before the age of Wikipedia, searching for answers in encyclopedias and looking up words in dictionaries. I like to write about things, people and places I wish I knew or I want to know better. If I held to the "write what you know" maxim, I wouldn't have written this book. I wasn't born or raised in Huntsville, Alabama. I knew very little about the South. I thought cotton was just something my mother stuck in my ear when I had an earache. Until recently, I believed a cotton gin was where moonshine was made. The first time I laid eyes on a cotton field was the first time I flew to Huntsville from Detroit, Michigan, in 1986. The landscape looked like a patchwork quilt of green mountains, red clay and white cotton fields that reminded me of the snow back home. I've lived in Alabama thirty years now. I have raised my two boys here. But does that mean I know its people and its culture? Does it qualify me to write a book about a very important part of your regional history? I don't know.

How well do any of us know our own histories? How many times have we wished we'd asked our deceased grandparents or parents more poignant questions about their childhoods? I come from a family of auto workers, but I couldn't tell you how one of them spent their days on the assembly line or in the factories. But I wasn't asked to write a book about my family or my history; I was asked to write one about the history of

Preface

the people of Huntsville, Alabama, the place that's been my home for over half of my life.

I've spent months digging through old newspapers and journals, interviewing people, going to reunions, driving all over Huntsville and learning things about this city and its people I never knew before. I found out that long before rockets were part of the skyscape of Huntsville, the water towers and smokestacks of its textile mills were the monoliths of industrial progress. There are few people left who can tell you what it means to be a carder, spinner, loomer or doffer or what it felt like to hear that morning whistle blow, to breathe the white cotton dust that hung in the air or holler over the noise of the various machines. There aren't many still with us who sank their teeth into the first slaw dawg at Mullins, spent a dime to ride the streetcar downtown or caught a Saturday matinee at the Center Theater. Fortunately, there are still a few souls who proudly wear the moniker of "linthead." I am happy to have heard some of their stories. To those who are no longer with us, well, I am happy some of you, like historian Sarah Huff Fisk, wrote what *you* knew. It's to the lintheads, both living and deceased, and to your families, that I dedicate this book.

ACKNOWLEDGEMENTS

Writing a book that covers over a one-hundred-year-long span takes a lot of assistance and support. I would like to thank the following list of people for their contributions in getting this book to publication.

For their help in research, I thank Alexis Stratch for the hours of digging she saved me and Susanna Leberman and the other librarians in the Heritage Room at the Huntsville–Madison County Public Library for pointing me in the right directions. For their hospitality and keen memories, I thank Geraldine Walker, Bill and Doris Brown, Maebelle Winkles, Betty Owens, June Golden, Larry Lyons, Arni Anderson and all of the former mill villagers and workers who shared their stories and photographs. For keeping the legacy of the lintheads alive, I thank Jim Bryne, Jim Hudson, Donna Castellano, Jessica White, Court Heller, Frances Akridge, Debra and Alan Jenkins, Dale Bowen, Elizabeth Tubbs, everyone at Lincoln Ministries, Rick McNully and his family, Ryan and Brittney Saffell, Mayor Tommy Battle and the many other government officials, community leaders and volunteers working to preserve and revitalize these historic sites. I thank my good friend Peggy Bilbro for her editing eye; my husband, Ray, for his support and technical savvy; my son, Logan Tanner, whose mural sparked my interest in the history of the textile mills; and my other friends and family members for their patience in listening to me ramble on about the facts and discoveries I've learned along the way. I hope all Huntsvillians enjoy the book and come away learning something about their city and its people that they didn't know before.

1
THE SOUTHERN MIGRATION OF NEW ENGLAND'S COTTON MILLS

Pre–Civil War New England

From "Cottage Industry" to Factory

Cotton, one of the first cultivated plants, has been part of human culture since prehistoric times. Archaeologists working in a cave near Mexico City unearthed bits of cotton bolls and pieces of woven fabric that proved to be at least seven thousand years old. As far back as 3,000 BC cotton was grown, spun and woven into cloth in the Indus River Valley of Pakistan.[1] The first written records concerning Indian cotton appeared in early Buddhist and Hindu texts.[2] Arab merchants brought cotton to Europe about AD 800. Hundreds of years before the Aztec, Mayan and Incan civilizations existed, cotton was cultivated in South America, the valley of Mexico and the Caribbean. It was grown by Native Americans in the early 1500s. The Spaniards raised a cotton crop in Florida in 1556.[3] The fluffy white bolls have always lent themselves admirably to commerce and would become a major source of agricultural and industrial growth in this nation.

During the late eighteenth century, the conversion of raw cotton into cloth was primarily a home-based business, with the work done by hand or with rudimentary equipment, such as the spinning wheel and hand loom. Entire families got involved with the work—men were often weavers, while children assisted in cleaning the raw cotton and women spun the materials into threads or yarns. British inventions, such as James Hargreaves's multi-spindle spinning jenny and Richard Arkwright's water frame, would

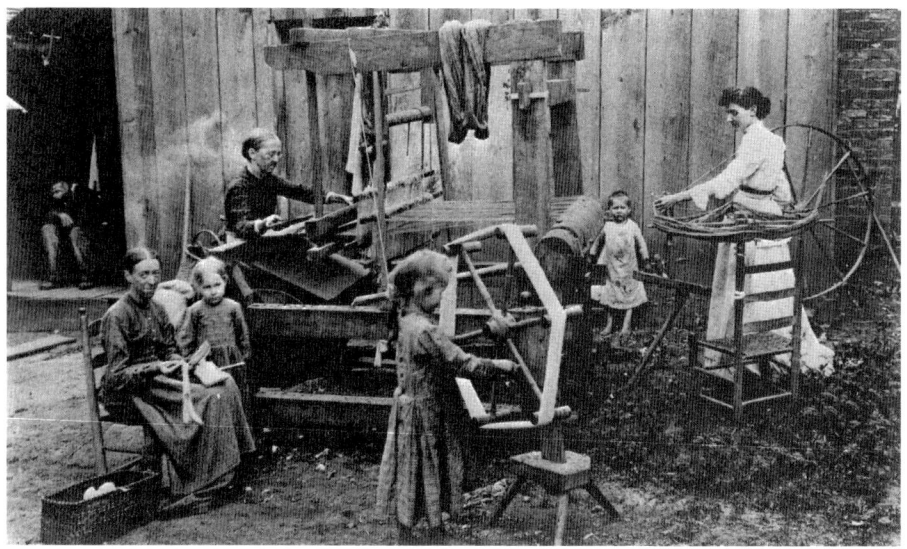

The processing of cotton was once a home-based cottage industry. *Huntsville–Madison County Public Library Archives.*

revolutionize the spinning process and move the one-time cottage industry into water-powered mills located by streams. Arkwright built the world's first successful water-powered cotton-spinning mill in Cromford, a village in Derbyshire, England, in 1771.[4] This was a vital step toward full-scale factory production and helped to launch Great Britain into an Industrial Revolution far ahead of the United States.

While the early colonists had the ability to produce much cotton, they lacked the equipment and mechanical know-how to process the raw product on any large scale. In eighteenth-century New England, the conversion of raw cotton into cloth was still a home-based business. Two prominent men helped to usher cotton manufacturing into this country: Samuel Slater and Francis Cabot Lowell.

Slater, a skilled British textile machinery engineer and apprentice of Arkwright's, came to Rhode Island in 1789 and developed the country's first cotton spinning mill. Lowell, a member of a prominent New England mercantile family, established the first integrated cotton spinning and weaving facility in what later became Lowell, Massachusetts.

British law did not permit trained mechanics to leave the country, but twenty-one-year-old Slater, dressed as a farm boy, boarded a vessel to New York with the designs of Arkwright's water frame squirreled away in his brain. He first landed in New York and went to work for the New York Manufacturing Company. Slater quickly became frustrated with the facility's inadequate machinery and spring-fed water supply. Upon learning of a wealthy Rhode Island mercantile family's search for a skilled textile manufacturing technician, the clever and confident young man promptly wrote Moses Brown touting his expertise. An agreement was reached with the Brown family, and by 1791, Slater had perfected a twenty-four-spindle water frame, based on Arkwright's design. In his hometown of Belper in Derbyshire, he was dubbed "Slater the Traitor," but Andrew Jackson later gave him the illustrious title of "father of American manufacturers."[5]

In 1793, a few miles from Providence, Rhode Island, on the banks of the Blackstone River, the firm of Almy, Brown and Slater hired local artisans and laborers to construct a wooden building suitable for manufacturing cotton thread by water power. Slater Mill was the first successful cotton spinning factory in the nation. Slater first hired children to do the work in the mill, as they provided cheap labor and had the nimble fingers necessary for the job. But as the mill grew, he realized that poor white families were an attractive labor pool and built housing to attract them. He aptly named this

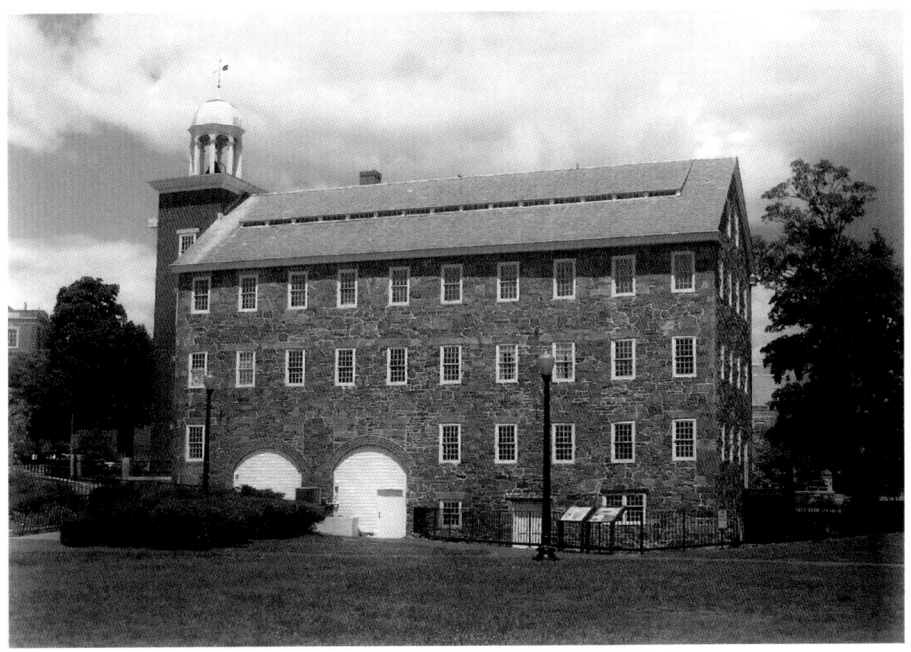

Slater Mill Historic Site, Pawtucket, Rhode Island. *Creative commons licensed (CC BY-SA 2.0); Flickr photo by Doug Kerr: https://www.flickr.com/photos/dougtone.*

first mill village Slatersville, a name the area still holds today. This worker-village system was one that would be replicated throughout the Northeast and, later, in the South.

In the early nineteenth century, twenty-five miles northwest of Boston, Harvard graduate Francis Cabot Lowell was making his own mark as an agent of the Industrial Revolution. Lowell came from a wealthy Boston merchant family. Like Slater, he "borrowed" British technology to further mechanization in the United States. On a trip to England in 1811, Lowell spied on the British textile industry, memorizing the workings of their power looms. Back in Boston, he re-created a loom that could leverage water power to mechanize both spinning and weaving so that the processes could now be done under the same roof. Lowell also developed a joint-stock company, selling shares to raise capital. His combination of innovative technology, novel organization design and creative financing proved extremely successful. By 1840, the city of Lowell had developed into the second largest in the state of Massachusetts, with nine large-scale mills backed by $8 million in capital.[6]

Slater's and Lowell's inventions sped up the processes of spinning and weaving and increased the demand for cotton. In 1794, Yale graduate Eli

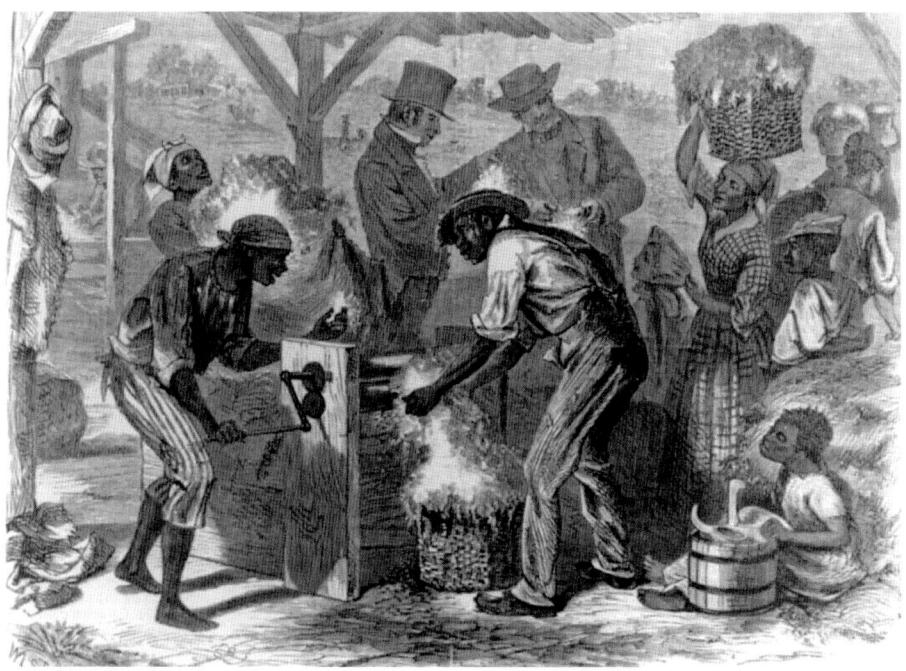

African American slaves using the first cotton gin, 1790–1800. Illustration in *Harper's Weekly*, December 8, 1869, drawn by William L. Sheppard. *Library of Congress*.

Whitney invented the cotton gin (short for engine), a machine that would transform the carding process—which separated the seed from the fluffy white bolls. In 1830, before Whitney's gin was in widespread use, the southern United States produced roughly 750,000 bales of cotton. By 1850, that amount had exploded to 2.85 million bales.[7] The faster processing of cotton meant it was profitable for southern landowners to establish larger cotton plantations and bigger cotton farms, which necessitated more slave labor. The slave population in the United States increased nearly five-fold in the first half of the nineteenth century. While it took a single slave about ten hours to separate one pound of fiber from the seeds, a team of two or three slaves using Whitney's cotton gin could produce around fifty pounds of cotton in just one day.[8] By 1860, the South was providing about two-thirds of the world's cotton supply. Southern wealth had become reliant on this one crop and thus was completely dependent on slave labor.

The Southern Antebellum Mills

Slaves and "Poor White" Workers

Most of the growth in manufacturing remained in the New England region, especially in Massachusetts, during the antebellum period. The southern states stayed overwhelmingly agricultural. The factory system made limited strides in the southern states pre–Civil War but was not nonexistent. In fact, the southern industry saw fairly impressive growth and profits during the 1840s. There were nineteen textile mills in production in Georgia in 1840.[9] In 1845, Virginian William Gregg built the South's first large-scale textile factory and the first mill village, Graniteville Mill, in Horse Creek Valley, South Carolina. Alabama remained the leading cotton producer in 1850, even as it continued to industrialize. Southern mill expansion continued, but at a much slower pace, through the remainder of the 1850s.

Mill owners at this time had essentially two worker pools to choose from: slaves and their families or the "poor white" rural families. Often it was not a matter of choice but of availability in a particular area. Many of the early mills, having been founded and run by plantation owners, were small and needed only ten to twenty workers, often women and children from the mill owners' own plantations. Although planters continued to own and control most of these smaller mills, some of the day-to-day operations in the larger factories passed to intermediaries. Skilled workers and supervisors from the North sometimes filled these roles of training and instilling "proper" work habits and values. Many of the white workers clung to their rural, preindustrial work habits and were used to setting their own pace, so these intermediaries had their work cut out for them.

By purchasing slaves, rather than hiring whites, factory owners hoped to avoid high turnover, save money and gain stability in their operations. Even Gregg, an apostle of white labor, conceded that slaves were suitable factory workers. He is noted as saying that slaves did not have their lives interrupted by schooling, did not suddenly quit and did not wander off to summer fishing holes. However, at his own factory, he employed predominately "poor, ignorant, degraded" white workers. The few slaves employed performed menial jobs outside of the factory.

The shift from black to white workers was neither immediate nor universal. While some mills employed all slave workers and some all white workers, others had an integrated workforce. However, the 1850s brought heightened racial consciousness and growing criticism from white workers about unfair competition with slave workers.

White workers, for the most part, earned a good wage and had educational opportunities, housing and land for gardening, but they failed to appreciate these benefits. As the mill owners expected, turnover was high. Regardless, the supply of poverty-stricken white workers was abundant, and operatives were easily replaced. Textile manufacturers appreciated the work ethic of slaves and may have wanted to employ them, but by the 1850s, they found it difficult to purchase or hire them. Along with the escalating racial conflict, cotton prices rose during this decade, and slaves became more valuable as agricultural laborers. Thus, slaves returned to the fields, and mill owners reluctantly accepted a white labor force in their place.

The Civil War

The Cotton Trade Continues

Just as textile mills started to flourish in the South, the Civil War began, and any industrial expansion begun during the antebellum period was stymied. The southern industrialists had lagged far behind their northern counterparts with regard to textile manufacturing, partially due to the fact

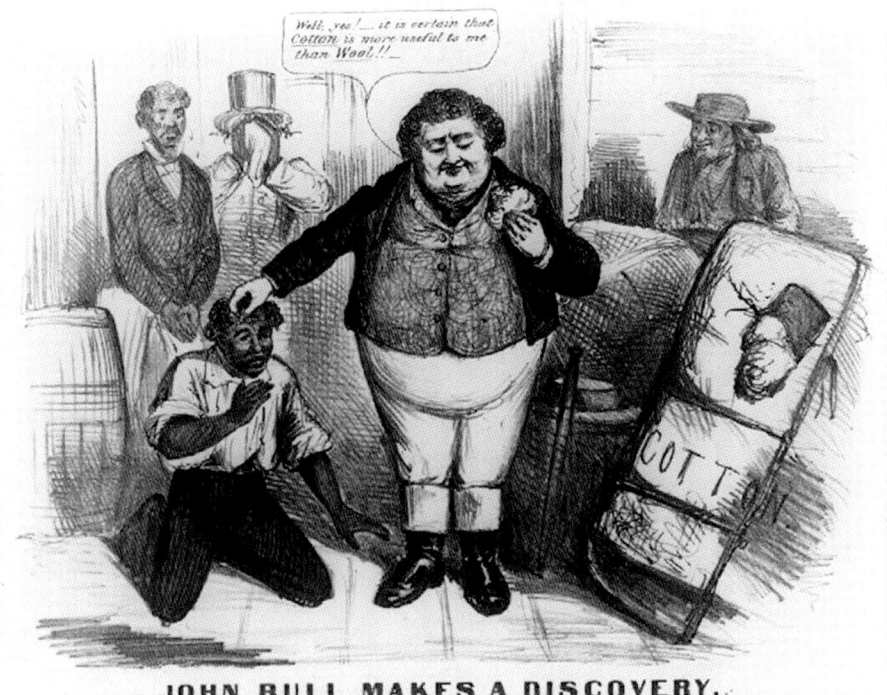

JOHN BULL MAKES A DISCOVERY.

An anti-British satire, reflecting Northern fears of English assistance to the Confederacy. *Library of Congress.*

that their money was tied up in the slave market and the cultivation and harvesting of cotton. By 1820, all of the northern states had outlawed slavery, but the reign of "King Cotton" played a major role in the growth of the slave economy of the South. Most high school textbooks would have one believe that the war was about the abolition of slavery, but ironically, without the slave workforce, New England and Great Britain textile manufacturers would not have their coveted cotton crop. Slavery, and thus the war, was in fact prolonged because of the dependence on cotton and the need to keep the slave workers out of the factories and in the fields. Eventually, this same war would prove to be a turning point in the rise of cotton manufacturing in the South.

What is also seldom mentioned in textbooks is that while the Union blockade separated cotton from its market, there remained some semi-illicit trade between the Union and the Confederacy throughout the war. By this time, the North had a ravenous demand for cotton, and the South needed provisions to feed and care for its many laborers.

In his article "Traders or Traitors: Northern Cotton Trading During the Civil War," author David G. Surdam gives three primary reasons why northerners desired to continue the cotton trade:

> *First, southerners and northerners alike thought that Europeans might intervene if their cotton textile manufacturers were deprived of raw cotton.... Second, many northerners, Lincoln included, believed that latent Unionism was strong throughout the Confederacy, but especially in the border states. To foster Unionism, Lincoln countenanced small amounts of trade to succor loyalist in occupied areas.... Third, northern textile manufacturers needed raw cotton to remain in business.*[10]

Profits derived from the trade in cotton had fueled the wealth not only of northern factory owners but also those in Great Britain. In fact, Liverpool was the most pro-Confederate place in the world outside of the Confederacy itself. Liverpool merchants helped bring out cotton from ports blockaded by the Union navy, built ships of war for the Confederacy and supplied the South with military equipment and credit. The British weekly the *Economist*, generally a strong opponent of slavery, nonetheless feared the prospect of abolition in the South: "The catastrophe would be so terrible, its accompaniments so shocking, and its results everywhere and in every way deplorable, that we most earnestly pray it may be averted."[11]

Even with Liverpool's aid, Union blockades were able to significantly reduce the volume of cotton exports. In 1864, exports were only one-eighth of the South's prewar volume.[12] The South's most powerful economic weapon during the war was cotton, and it was not all that eager to let it go. When the Union started its blockade in 1861–62, Confederate leaders "obsessed by the notion that cotton is king, welcomed it almost as a godsend."[13] They wanted the British on their side, and cotton was the dangling carrot. Yet little did they know that Britain, having anticipated the war, had long been stockpiling cotton in order to avoid dependence on the American South and, thus, owed them no alliance unless the war prolonged, which of course it did.[14]

To further minimize any gains the Confederacy might derive from trade, the Northern administration devised a system in which treasury officials would pay three-fourths of the current market price for cotton. The provisions the South received in turn were sometimes less than one-third the value of the raw product.[15]

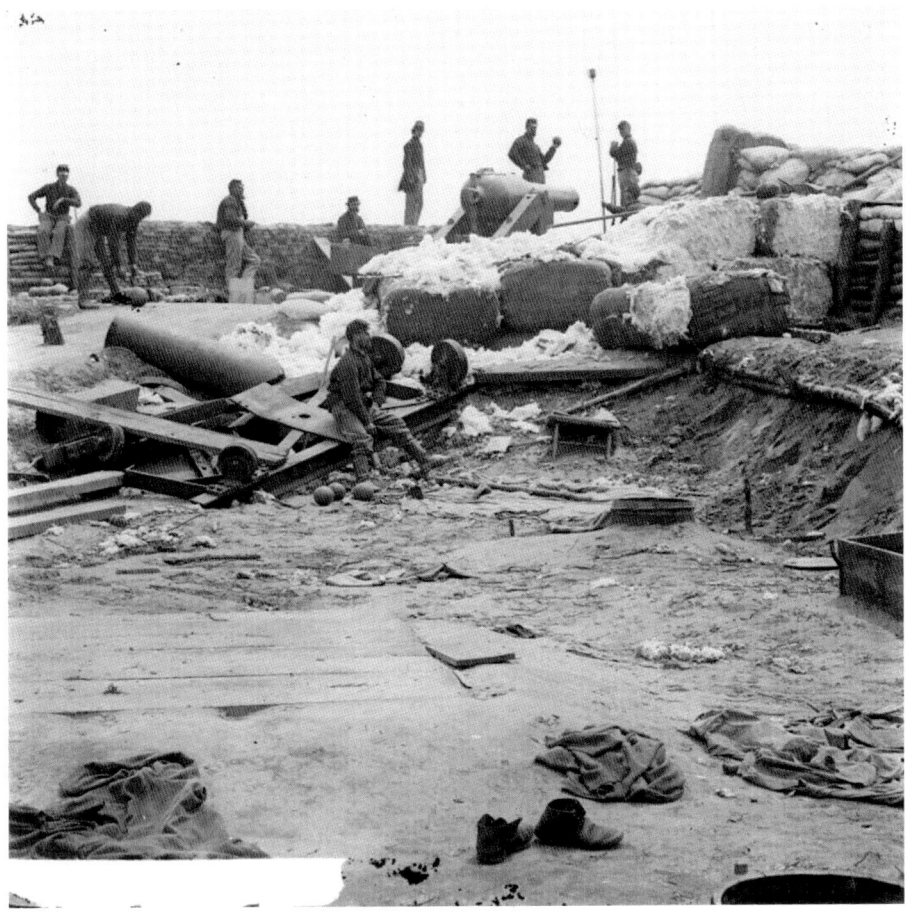

Yorktown, Virginia Confederate fortifications reinforced with bales of cotton, June 1862. *Library of Congress.*

Lincoln was adamant that trade be open to all loyal citizens. Traders had to apply to the administration, and the president made the final decisions on who filled the positions. Treasury agents were assigned to investigate the trustworthiness of trade applicants. Of course, any merchant who could cheaply transport cotton and goods stood to make a handsome profit, and greed became the driving force of the trade system. Prominent administration officials, politicians and businessmen all lobbied to have their friends, family members and associates appointed. Lincoln cast a blind eye to any misdoings under the belief that all was being done for the good of the Union and the nation.

Many military officers were opposed to the trade. Generals William Sherman, Ulysses Grant and Edward Canby were staunch opponents. Jefferson Davis was ambivalent and never truly embraced the necessity of it. So strongly did the Confederacy believe in cotton's sovereign power that the Confederate Congress required landowners to "destroy all cotton that might fall into the hands of the enemy." An estimated 2.5 million bales of cotton were destroyed by their owners or the Confederate armies.[16] Nevertheless, a total ban on trade would have generated a virtual black market on cotton and only magnified its profits to Northern traders. The inequitable trade arrangement did little to benefit the South and only prolonged the war. Most cotton farmers were left destitute by war's end. For the farmers and former slaves alike, the Reconstruction period would be fraught with struggle.

Post–Civil War South

From Field to Factory

The abolition of slavery brought about a major reorganization of agriculture. Impoverished cotton plantation owners, needing to recoup their losses, divided their land into smaller plots to be tended by individual black and poor migrant white families who were paid a share of the crop. The strenuous work and hot sun in the cotton fields took their toll on all workers, black or white, as author Elise Hopkins Stephens illustrates vividly in her book *Historic Huntsville: A City of New Beginnings*:

> *Between 1865 and 1900, in their struggle to make ends meet, these once proud people became bent over, their backs to the sun. The image of slave cotton pickers still bearing their cotton crosses comes to mind. Only now the faces weren't all black, but white too. And the burn of the sun made "rednecks" of many of them.*[17]

Local merchants set up crop-lien systems. Between the end of the Civil War and the 1930s, southern cotton farmers relied on the crop-lien system for credit so that they could survive until the crop came in each year. Tenant farmers and sharecroppers who were not landowners had to get food and supplies on credit from local business owners. When the crop came in, these merchants had a lien on it, and they received the first share of the profit,

Linthead Legacy

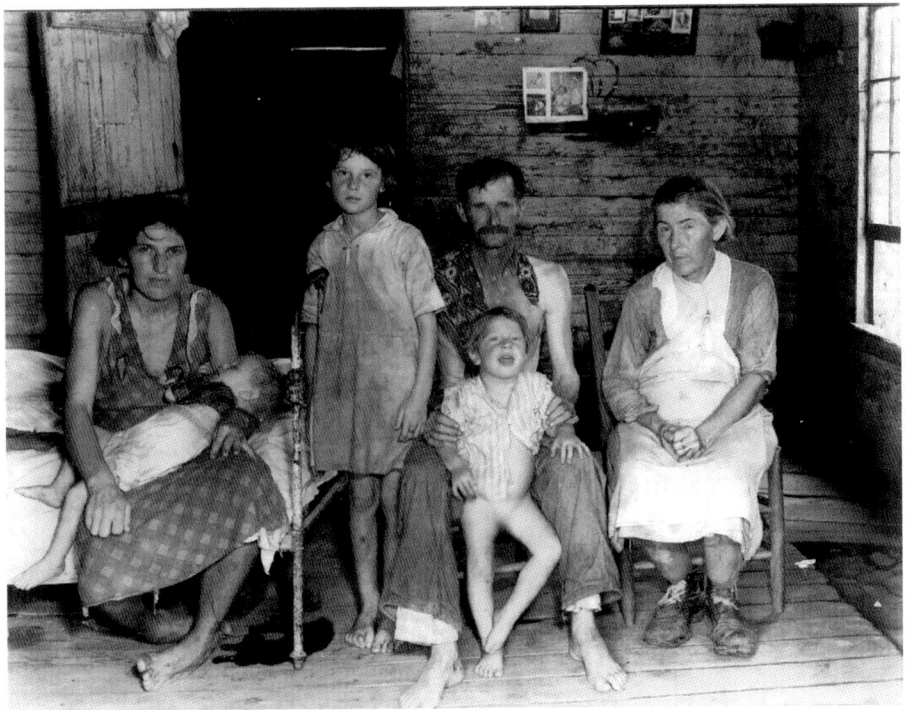

Sharecropper Bud Fields and his family at home in Hale County, Alabama, 1935–36. Photographed by Walker Evans. *Library of Congress.*

with the leftovers—meager as they might be—going to the farmer. To ensure payment, the merchant often stipulated the farmer plant a cash crop, such as cotton. Sharecroppers often ended the year in debt to both the landowner and the credit merchant. Agriculture during this time was almost totally invested in the cultivation of cotton, stifling any other industry that might have developed. For at least two generations after the American Civil War, the South remained predominantly invested in cotton cultivation and largely outside the industrial expansion of the Reconstruction period.

An event meant to accelerate and improve cotton production inadvertently sparked the southern redevelopment of the textile industry that had been halted by the Civil War. Edward Atkinson, a textile executive, banker and entrepreneur from Boston, hoped the 1881 Textile Exposition in Atlanta would heal the wounds left by the war and strengthen the relationship of the North and South with regards to cotton cultivation and manufacturing. The goal was to persuade farmers to refine their methods of growing, ginning and baling for shipment to northern mills. What better way to do

HUNTSVILLE TEXTILE MILLS & VILLAGES

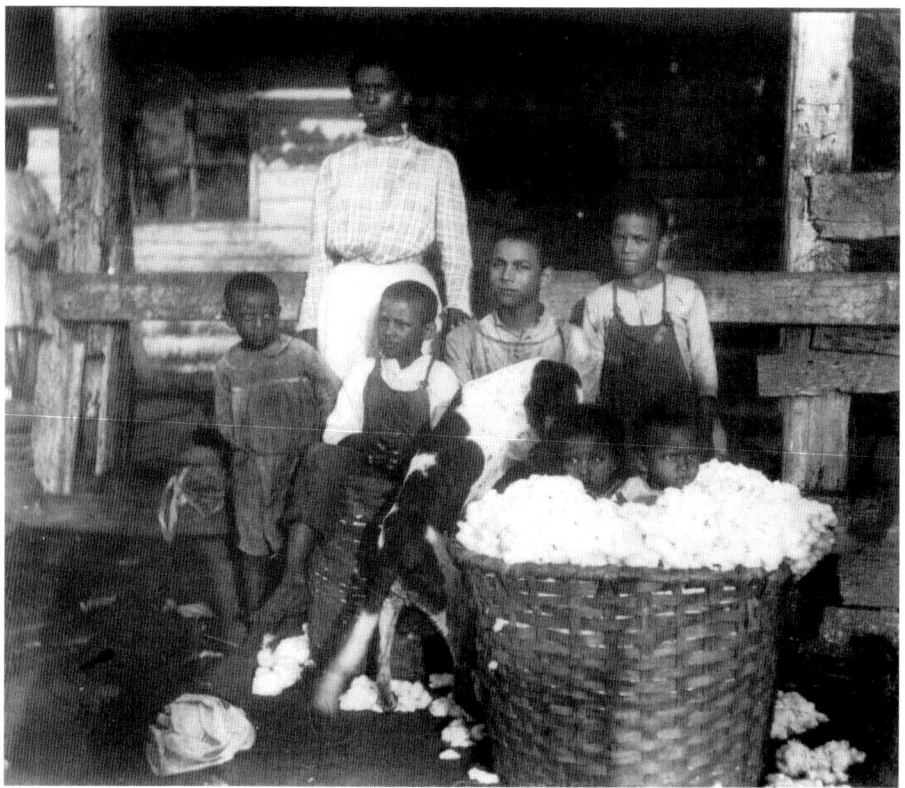

Black sharecropper family. *Courtesy of Larry Lyons.*

this, he thought, than to include them by demonstrating the processes used to take their raw product and turn it into cloth. Textile moguls, engineers and bankers in both the North and South financed the big event. The main exposition building was designed as a state-of-the-art cotton mill containing the latest in equipment, with the ultimate goal being it operate as an actual mill at the close of the exposition. In 1882, that intention became a reality thanks to the investments of a number of local businessmen in the manufacturing enterprise.[18] The objectives of the northern capitalists had backfired, however, and ambitious businessmen of the postwar South realized money could be made by bringing the "cotton mills to the fields," a slogan of the resulting southern mill campaign.[19]

In the late nineteenth century, merchants with money to invest were beginning to back mill construction. Southern textile promoters generally came from the southern business class, with personal or family backgrounds in banking, railroads and commerce. Railroad expansion was tantamount to not only the growth of commercial agriculture but also the development of towns and, ultimately, the construction of cotton mills. The southward migration of production broadened during the mid-1890s. While many mills were locally owned and financed, a number of major New England corporations, including the Merrimack Manufacturing Company of Lowell, Massachusetts, built southern facilities to manufacture goods.

The growth of the industry created new job opportunities for indigent farm families. In factories that had both spinning and weaving capacities, there were a variety of positions available. These various jobs, with such unusual names as "slubbers" and "doffers," are described aptly by historian Whitney Adrienne Snow in an article in the *Alabama Review*:

> *When cotton arrived at the mill, pickers unwrapped the bales and removed unwanted materials. Lappers straightened the fibers while slubbers lessened the size of and interwove the fibers. Bobbers placed rolls or spools of roving in creels or bars that kept the bobbins on spinning machines and ran the roving between two rollers operating at different speeds. Doffers replaced full bobbins. Spinners tied together loose ends and often operated a number of spinning frames, machines that stretched and twisted fibers, carders placed fibers between toothed cards that paralleled and straightened the fibers. Warpers either twisted the yarn into string or made it into rope. Weavers operated looms that wove fabric. Slashers added starch and tallow to strengthen the product, drew thread between a heated cylinder, and wound the ends on loom beams.*[20]

From the 1880s through the Great Depression, thousands of poor white farmers traded the fields for the factories. Southern mill owners hired entire families, including children, paying less than a living wage. Many mill promoters believed that by giving the farmers jobs they were not only freeing them from poverty but teaching them the merits of frugality, punctuality and obedience. Slave operatives had been common in the antebellum mills, but the postwar mills were a place where the black man had almost no place. The free black man, if employed by the mill, performed some of the rough outside work, tending the pickers and baling, or clean-up work such as sweeping or scrubbing. Rarely were mill jobs integrated.

Along with the towns and factories, mill villages were built to house families, with schools, stores and other facilities. The presence of the mill was regarded by residents as a source of civic pride and a symbol of regional

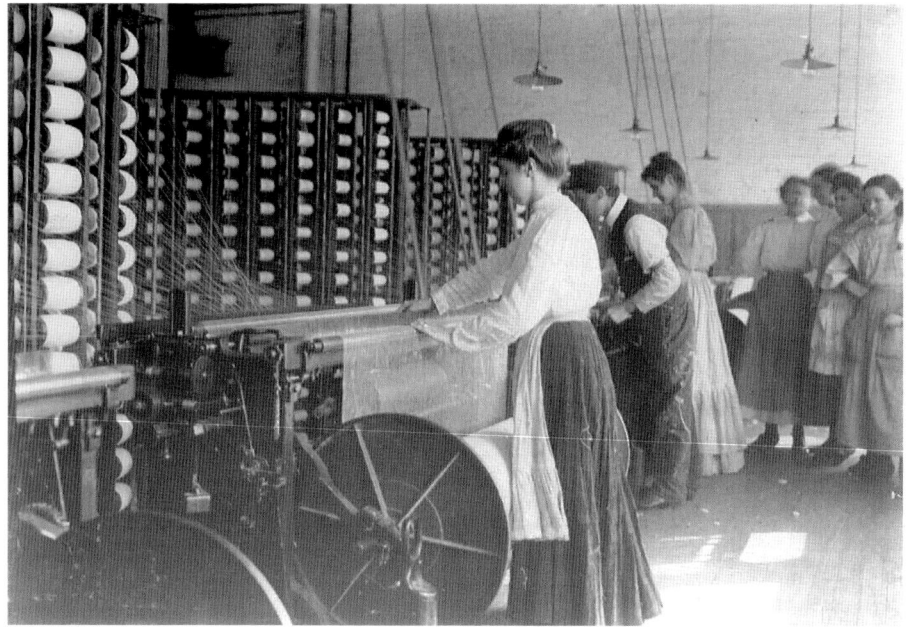

Mollahan Mill, Newberry, South Carolina. Young woman at spinning machine, date unknown. Photo by Lewis Hine. *Library of Congress.*

regeneration, even if the workers themselves were deemed "white trash" and given the moniker "lintheads" by the city dwellers because of the white cotton dust that covered their heads and clothing. The millworker/manager relationship was a paternalistic one wrought with hardship. Much as the sharecropper had been dependent upon the landowner and merchant, the millworker became dependent upon the mill owners and managers.

The southern and northern mills remained competitive throughout the early 1900s, with the southern cotton factories producing a widening range of quality goods. The lower wages, longer work hours and cheap labor source in the South helped to give southern manufacturers a competitive advantage, and by 1925, the South finally surpassed New England in the number of spindles in operation.[21]

2
THE FIRST HUNTSVILLE COTTON MILLS

Cabaniss Spinning Mill

In 1809, a full decade before Alabama would become a state and when the textile industry in New England was still in its infancy, a Tennessee contractor named Charles Cabaniss, with the help of engineer C.P. Poole, began designs on the first spinning mill in the South. The spot Cabaniss chose for his mill was the Barren Fork of the Flint River, twelve miles northeast of Huntsville (then known as Twickenham). This area, where two streams of water converge and become the single, continuous flow of the river known as Three Forks, was an excellent water power site. The Cabaniss Mill was completed in 1815. There are few accurate records left about this factory; most have been pieced together by Dwight M. Wilhelm in his book *History of the Cotton Textile Industry of Alabama: 1809 to 1950*.[22]

Haughton Mill and the Bell Factory

On September 4, 1819, Horatio Jones formed a cotton-spinning factory on the Flint River and began producing slave clothing. By 1823, the company had dissolved due to financial problems. Jones, however, refused to give up and soon formed a new endeavor. On October 21, he announced his plan to once again spin cotton and coarse shirting and moved the factory downstream. That mill, too, proved unsuccessful, and later that year,

Jones sold the company to a North Carolinian, William Haughton. An advertisement in the (Huntsville) *Alabama Republican*, dated September 29, 1820, is one of the few records found regarding Haughton's Mill:

> "COTTON FACTORY"
>
> *The proprietors of the Cotton Factory at Haughton's Mill, near the Three Forks of Flint River, would inform the inhabitants of Madison and adjoining counties, that they have made some additions to their machinery, and have constantly on hand an assortment of spun cotton, which they will exchange for good, clean seed cotton on accommodating terms.*
>
> *—Horatio Jones & Company.*

In 1829, Haughton chartered the mill to a group of individuals, including William Patton and James J. Donegan. In December 1832, the mill—renamed the Bell Factory—was incorporated by the General Assembly for $100,000. But the property itself was not purchased until two years later, on April 3, 1834, when Patton, Donegan and Company gave the sum of $20,000 for the mill and one hundred acres of land to some of the other men from the original charter. It was operated for the most part by Patton, Donegan and Company, with C.P. Cabaniss, the son of Charles Cabaniss, later becoming affiliated. William B. Tabor was one of the earliest superintendents of the factory. W.H. Echols was secretary/treasurer and, in the factory's later years, served as superintendent. In addition to the three-and-a-half-story mill, a store and warehouse were maintained in Huntsville on the Square.[23]

The Bell Factory was the first spinning and weaving factory in Alabama. The power used in the operation of the three thousand spindles and one hundred looms was furnished by damming the water of the Flint River and forcing it over a wheel. The mill would not become steam powered until 1868; therefore, in lieu of a steam whistle, the slave laborers were called to work by the ringing of a large bell, hence the name. The factory workers lived in thirty-eight cabins organized by household inside the factory compound, which was surrounded by a wall. The wall had a night watch and was closed to outsiders. While the workers were confined, they were not usually mistreated, as their hard work and loyalty were essential to the profitability of the business.

Some of the larger mills employed northern-born supervisors to train and guide their workers, but such employees were too expensive for smaller

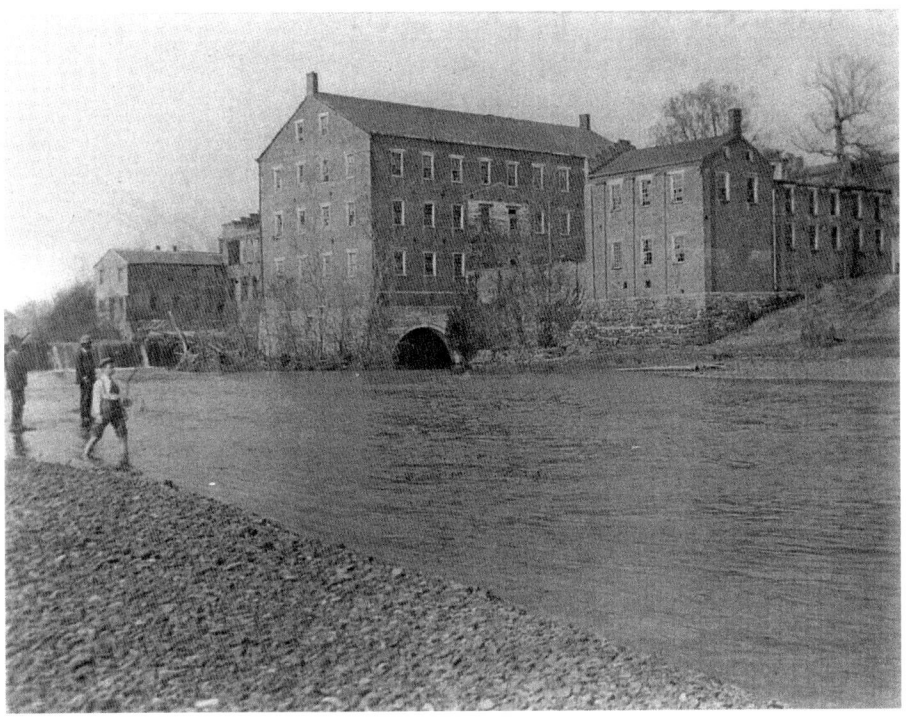

The Bell Factory, built in 1820. The original building burned in 1841. This is the stone and brick mill built as its replacement the following year. *Huntsville–Madison County Public Library Archives.*

manufacturers, which sometimes entrusted talented slaves to such duties. At Bell Factory, a slave named Branch managed forty hands in the spinning room on the third floor. He was the middleman, dispensing the master's orders and acting as spokesperson for the workers. This is not to say everything always ran smoothly. Bondsmen toiled long hours in tight spaces to produce the sheeting, plaid, ticking and yarn. Their fatigue and lack of freedom sometimes fostered discontent. Thievery was a problem. Some disgruntled workers took cuttings and scraps for their own use or exchanged stolen raw cotton and cloth for food and other items. Frustration and resentment could also take the form of arson. In 1841, the Bell Factory burned to the ground. A slave was rumored to have set fire to the mill, though other sources cite the cause of the fire as a mystery.[24]

By June of the following year, Patton, Donegan and Company had rebuilt the mill and even installed an auxiliary steam plant. The factory ceased operations for a brief time during the Civil War. In 1868, it became the Bell

Factory Manufacturing Company and transitioned totally to steam power. While prior to the war the operatives were exclusively slave laborers, now mainly white women and children were employed, working twelve-hour shifts five days a week and earning only between eight and twenty dollars a month. Instead of cabins and a wall, there was now a mill village of nearly three hundred residents. In 1881, a visitor to the mill described the mill and mill village as follows:

> *The factory is large and roomy for the machinery, and everything seems more cozy and comfortable than in Northern factories....Each family has a house on the land of the corporation, a large garden, and a cow....In every home I saw a sewing machine. All have open fire places. Major Echols was evidently regarded as a friend by the families on whom we called.... No liquor is sold except under his direction. There are a church and a school....I thought it seemed a happy little community.*[25]

The overworked and underpaid but still generally happy workforce helped to make the Bell Factory highly successful. In the Alabama Manual and Statistical Register for 1869, Robert Patton, the twentieth governor of Alabama and son of William Patton, noted one extraordinary accomplishment of the manufacturer: "The Bell Factory, near Huntsville, in Madison County, is one successful operation....Specimens of the Bell Factory were forwarded to the Paris Exposition, and in the report of the Committee, honorable mention is made of them."

Business continued until 1885, when more efficient equipment and other economic factors caused the mill to be abandoned. On October 27, 1957, a marker placed by Judge David Campbell and the local chapter of the Daughters of American Colonists was unveiled. It stands approximately ten miles northeast of Huntsville on Winchester Road. The factory bell—all that remains of the original building—was given to the Oliver

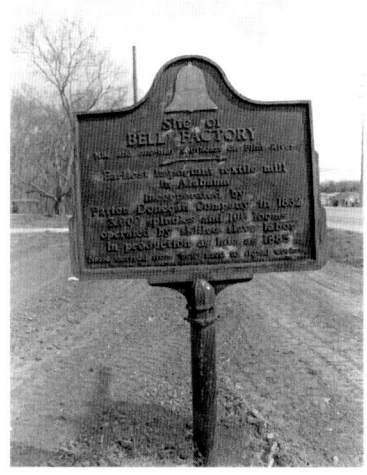

Bell Factory, the earliest important textile mill in Alabama. The building no longer exists, but this commemorative sign stands in the midst of progress of another kind—road construction. *Author photo.*

Bierne Patton Academy in 1910. Riverton School received the bell in 1918, and it remained there until 1973. The bell was placed into a tower in front of the new Riverton School building on Winchester Road and dedicated in 1975. The bell holds many memories for people in the community. Alice William Lay reminisced, "How well I remember hearing the ringing of the bell on the morning of November 11, 1918, when Dad and my brothers went to the school at 4:00 a.m. and rang it, spreading the news that World War I had ended."[26]

Huntsville Cotton Mills and West Huntsville Cotton Mills

By the 1880s, northern Alabama had begun to recover from the destruction left behind by the Civil War and the Reconstruction years. Leadership in the area actively sought capital for the manufacture of thread and yarn. Industrialization gained momentum in Huntsville and Madison County with the completion of the Nashville, Chattanooga and St. Louis Railway. Northern and western capitalists invested in real estate, and the processing of cotton expanded.

The very first mill within the city was the Huntsville Cotton Mills, built in 1881, a spinning mill near the Memphis and Charleston Depot on Church Street. D.L. Love of Mississippi, who recognized the opportunities in Huntsville, advertised in a Providence, Rhode Island newspaper and received a response from Joshua Coons. A trained cotton spinner, Coons not only traveled from Rhode Island to Huntsville to approve a mill site but also found directors and officers and persuaded the city to waive the mill's taxes for a decade.[27] By 1887, the mill had twenty thousand spindles. In 1918, it reincorporated as Margaret Mills and, in 1932, became Fletcher Mills. The Huntsville Cotton Mill was a predecessor of the larger mills to come.

Huntsvillians may know the surnames of Pratt, Wellman, Wells and O'Shaugnessy as local street names, but the names originally belonged to men who were instrumental in bringing the cotton mill industry to the city. Minnesotan Tracy W. Pratt and South Dakotan Willard I. Wellman met and did business together in Pierre, South Dakota. Pratt traveled to New York City to sell South Dakotan school bonds and there met James O'Shaughnessy. O'Shaughnessy convinced Pratt of the potential growth opportunities from Huntsville's cotton fields and workforce. James's brother,

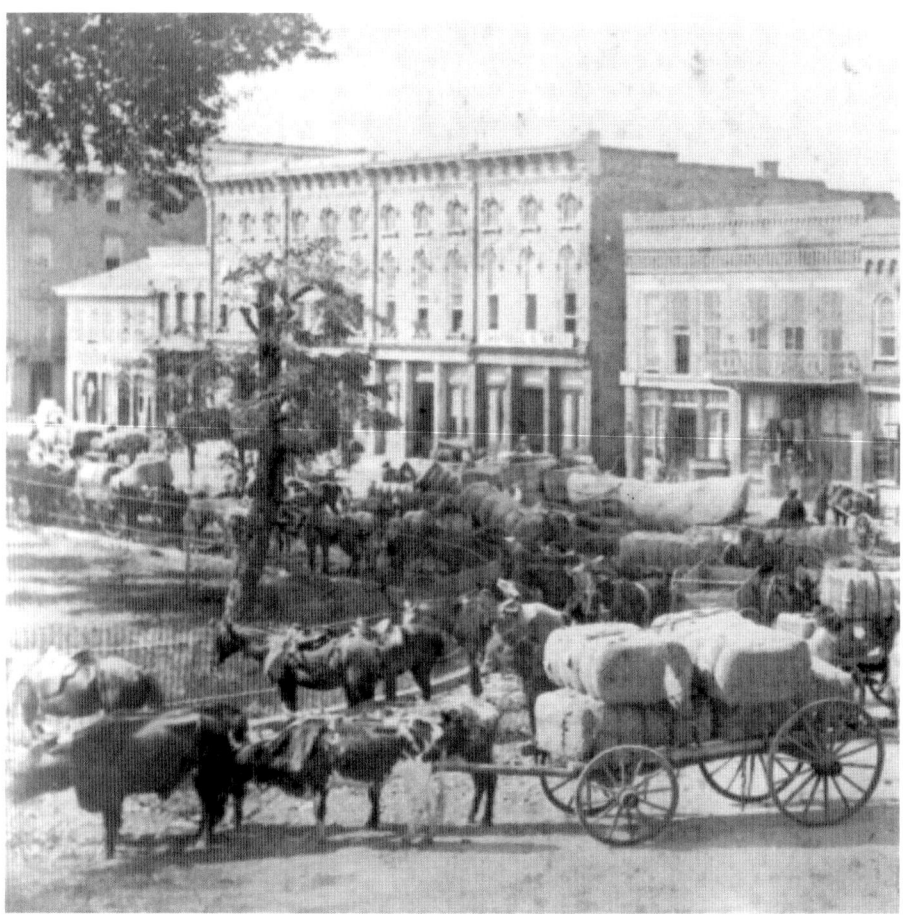

Cotton farmers and cotton merchants conducting business on "Cotton Row," downtown Huntsville, late 1800s. *Courtesy of Larry Lyons.*

Michael, had already purchased the abandoned machine shops of the Memphis and Charleston Railroad in the city and opened the Huntsville Cotton Oil Mill there in 1881. Michael built his family residence, a Queen Anne–style mansion, just off Oakwood Road and named it Kildare, after the county in Ireland where he was born. James owned several hundred acres on Monte Sano and purchased his family's summer home there. The O'Shaughnessys' business interests reached from Florida to Georgia and even into Nicaragua, where they invested heavily in the digging of a canal to aid their shipping endeavors. (When the Spanish-American War broke out, after which the United States decided to go with the Panama Canal,

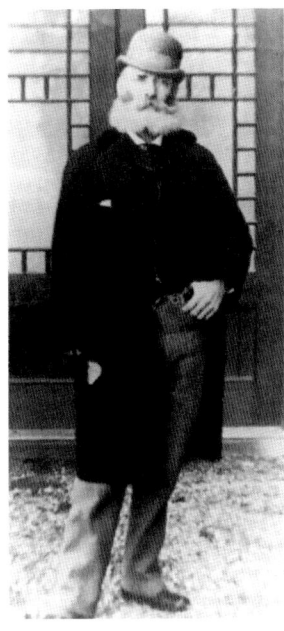

Michael O'Shaughnessy. *Huntsville–Madison County Public Library Archives.*

O'Shaughnessy lost what he had invested in this endeavor.) According to local lore, Huntsvillians never referred to them as carpetbaggers because "there wasn't a carpetbag made that could hold all of their deeds and securities."[28]

In 1886, the O'Shaughnessy brothers, eighteen local citizens and two men from Memphis organized the North Alabama Improvement Company (NAIC).[29] The enterprise culled the bulk of its membership not from politicians, bankers and outside developers but from the community as a whole, including physicians, grocers, an undertaker and a confectioner. James O'Shaughnessy described to Pratt the company's development projects, which included a hotel on Monte Sano. Pratt returned to Pierre to persuade his friend Wellman and another associate, William S. Wells, to make the move to Alabama. Another source says Wells had previously visited Huntsville on two occasions and met with James O'Shaughnessy himself. Whatever the case, the South Dakotans brought their Yankee know-how to the South sometime around 1892. A few months after his arrival, Pratt announced the formation of Huntsville's third cotton mill, the West Huntsville Cotton Mills, often called the Coons & Pratt Mills, on the corner of Ninth Avenue and Eighth Street. The mill was steam powered and had 5,200 spindles. Because he probably knew very little about the operation of a cotton mill, Pratt enlisted Coons, the one-time superintendent of the Huntsville Cotton Mills, to oversee daily operations. Aside from the mill structure itself, Pratt and Coons provided housing and stores for the workers.

Only two years after West Huntsville Cotton Mills started operations, northern manufactures began to feel the strains of competition, as indicated in an article in the *Florence Times* on March 31, 1894:

> *The West Huntsville Cotton Mills Invade New England*
>
> *Yesterday, Messrs Coons and Pratt filled an order of 70,000 lbs. of yarn to a Boston firm over all their eastern competitors. The order is not a large*

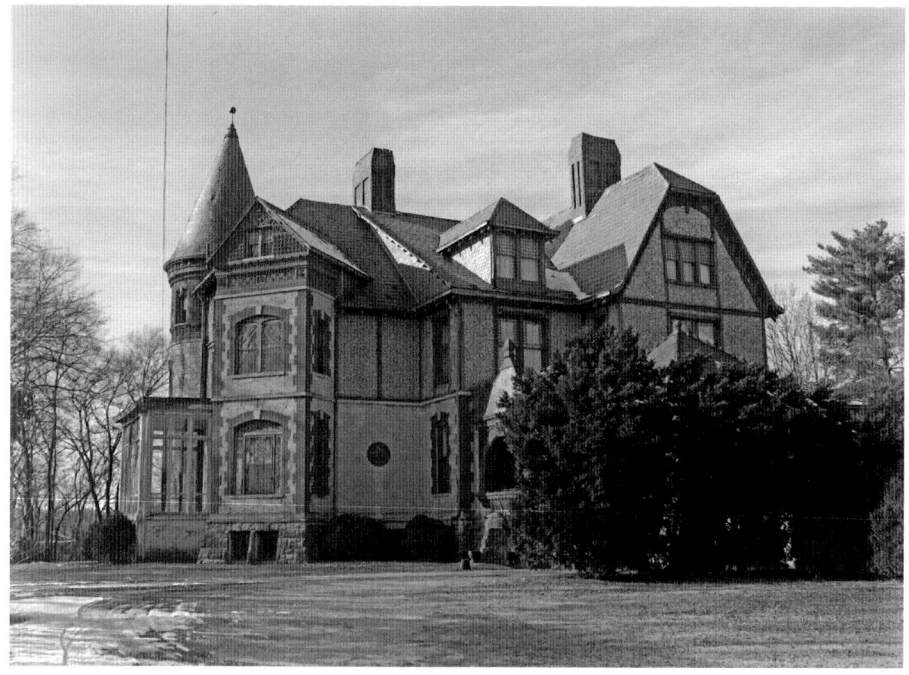

The Kildare-McCormick House, one of the state's finest examples of Queen Anne architecture. The structure is now, unfortunately, in a major state of disrepair. *Spyder_Monkey (Own work) [CC BY-SA 3.0 (http://creativecommons.org/licenses/by-sa/3.0)], via Wikimedia Commons.*

> *one, but the fact remains that our cotton mills are dangerous competitors for the New England spinners to buck against.*

In 1900, Tracy Pratt and his brother, Russell, purchased Coons's interest in the mill and made enlargements and improvements to the plant, as was alluded to in this article in the *Daily Mercury*, dated February 25 of that year:

> *The deal was closed a day or two ago but was kept quiet. The consideration which ran well up into the tens of thousand, has been paid over....The mill has 5280 spindles and has enough orders ahead to run it night and day for several months....Russell Pratt will take active interest in the management of the industry and will remove his family to the city from Minnesota at once. What the plans of Pratt Bros. are for the future is not given out but it is probable that the mill will be increased in capacity during the present season.*

Linthead Legacy

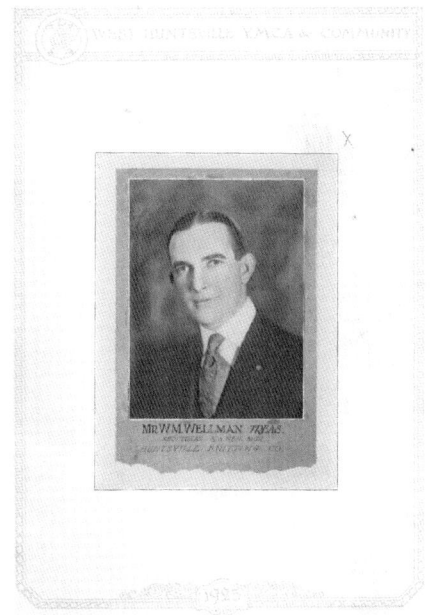 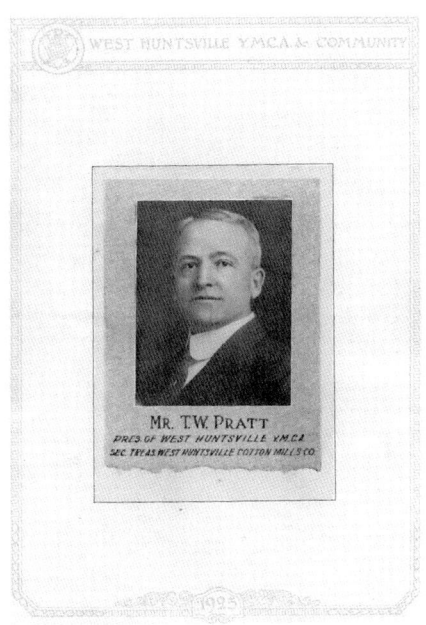

Left: Willard I. Wellman, treasurer of the West Huntsville YMCA and secretary-treasurer and general manager of the Huntsville Knitting Company, 1925. *Huntsville–Madison County Public Library Archives.*

Right: Tracy W. Pratt, president of the West Huntsville YMCA and secretary-treasurer of the West Huntsville Cotton Mills Company. *Huntsville–Madison County Public Library Archives.*

Growth in the industry continued, and by the early 1900s, there were nine textile mills in Huntsville. West Huntsville Cotton Mills was productive for the next thirty years.[30] By the time the tenth—Dallas Mill—opened, it was reported that Huntsville had more spindles than any other city in the South and was second in the nation only to Lowell, Massachusetts.[31]

Four of the largest, most successful and longest-running textile mills in Huntsville were Dallas, Merrimack, Lincoln and Lowe Mills. The legacy left behind by the workers who made these manufacturers profitable and the families of the mill villages is still felt within those communities today.

3
DALLAS MILL

The Factory

In 1890, the NAIC donated fifty acres of land in northeast Huntsville to induce Trevanion Barlow (T.B.) Dallas, a Nashville mill executive, to construct a cotton mill. "The mention of the name of Mr. T.B. Dallas in the cotton market and throughout the county," stated an article in the *Weekly Mercury*, "carries with it a prestige and confidence enjoyed in a great degree by none in the South."[32] Dallas was also lured by the city's board of mayor and aldermen, which passed an ordinance giving him 500,000 gallons of water for a term of ten years and an exemption from taxation.[33] He was persuaded by the locale, the kind words and most certainly the additional perks. In the spring of 1891, a Boston architectural firm, Lockwood and Green, began construction. Dallas and his partner, G.M. Fogg, also of Nashville, along with S.M. Milliken of New York, were major subscribers, along with the NAIC, the O'Shaughnessys and several loyal Huntsvillians, who supported the new industry to the extent of their constrained resources. The Milliken family supplied at least two presidents and one vice president for the corporation and maintained its interests to the end. Deering-Milliken Company of New York became the sole marketing agent for the mill.

In 1892, as the mill was nearing completion, a group of northern industrialists arrived in Huntsville and formed the Northwestern Land Association (NLA), which took up the cause of promoting Huntsville from the insolvent NAIC. Among the company's leaders were Wells, Pratt, Wellman and James A. Ward.[34] In May of that same year, the *Weekly Mercury*

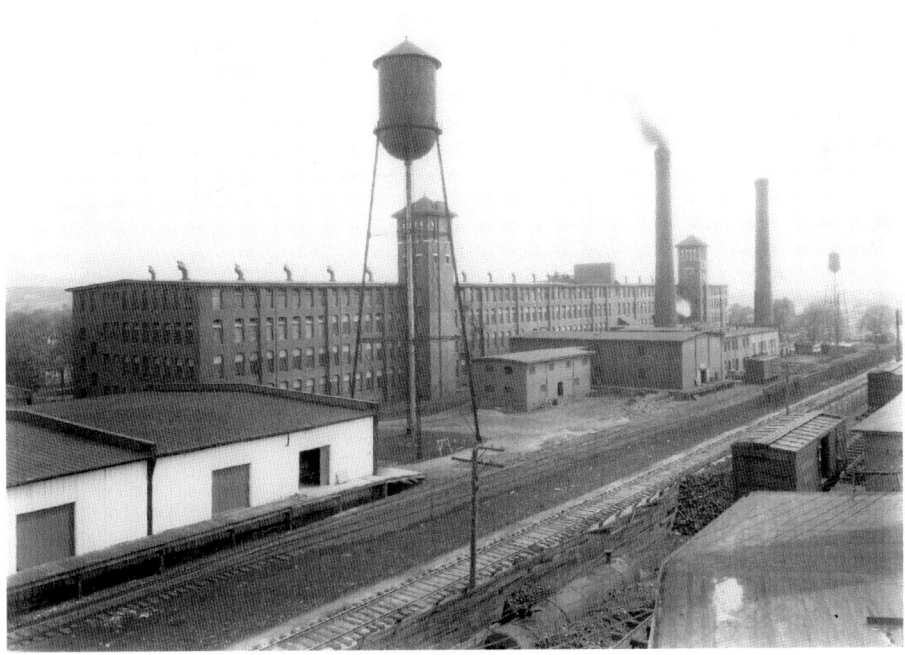

Dallas Mill, 1920s. *Huntsville-Madison County Public Library Archives.*

printed a rather turgid article boasting of Huntsville's suitability for the expanding industry: "We claim for reasons obvious to the intelligence of the world that this section offers greater inducement in the character of growing and ripening wealth for the home seeker and investor than any section now known in the United States."[35]

Dallas Mill began operation on November 16, 1892, with Fogg as its first president. The main structure was two football fields long and five stories high. In addition, there were several one-story buildings on the property. The mill manufactured bleached and brown shirting and the widest sheeting made in America at the time.[36] Originally steam powered, the facility later converted to electric power. The well in the basement of the main building tapped one of the main feeder channels of Huntsville's fifty-million-gallon-a-day "Big Spring," which supplied the power source.[37] In 1899, a 350-foot addition was added. At its peak, the mill employed 1,200 people and had 50,000 spindles and 1,541 looms. Dallas Mill grew to be the largest in the South.

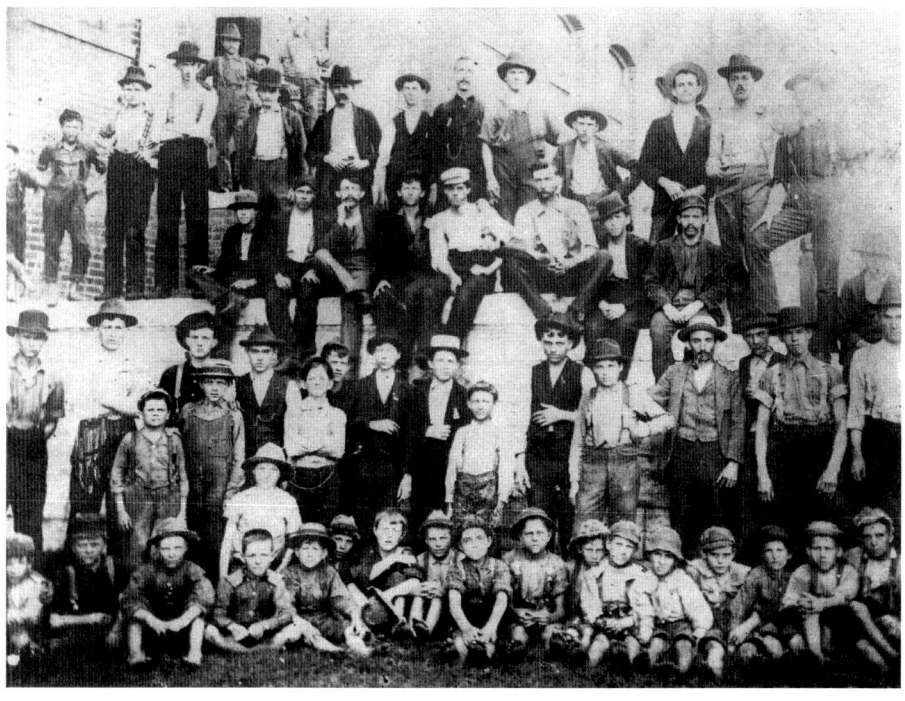

Dallas Mill loom fixers and doffers, 1898. Ten- to fifteen-year-old boys worked eleven-hour days for twenty-five cents a day. Men fared little better at seventy-five cents per day. *Huntsville–Madison County Public Library Archives.*

Soon after completion of the second mill, and during the first phase of housing development, the company was beset with scandal. In July 1900, Elijah Clark, a twenty-year-old black man, allegedly raped a thirteen-year-old girl who worked at the mill. An angry mob of one thousand millworkers left their jobs and stormed the jail. They lynched Clark in front of a crowd estimated to have swelled to six thousand before he could have a trial. His body was riddled with bullets. Dallas Manufacturing acknowledged that its employees had carried out the lynching and agreed to repair the damage done to the jail. Sadly, such upheavals were not uncommon at that time— workers returned to their jobs and calm returned to the village in short order.

T.B. Dallas served as general manager until his death in 1902. Like the O'Shaughnessys, Dallas died in debt. His family in Nashville had outlived their means, and as Dallas had been one of the biggest stockholders when the mill opened in 1892, his entire holdings were sold shortly after his death. His home on Monte Sano sold to William R. and Archie L. Rison, who both succeeded him as general managers of the mill. Dallas's home had previously been in the Rison family—long-standing members of Huntsville society—for many years.[38] Wells, Wellman and Pratt also made Huntsville their permanent residence.

Dallas Mill remained in full operation until 1949, except for periods when it was closed by labor union strikes in 1934 and 1947. During the 1947 upheaval, 650 workers—members of the Textile Workers of America—walked off their jobs. A mediation board gave up on a negotiated settlement in July 1948, and stockholders voted to sell the once immensely successful mill. In July 1949, the mill's machinery, real estate, private dwellings and buildings were auctioned off bit by bit. Most of the homes were sold to employees at reasonable prices.[39] Still, for over three hundred mill families—even those who were able to stay in their homes—the auction signaled an end to a way of life.

In 1955, two businessmen from Boaz purchased the mill and leased it to General Shoe Company (renamed Genesco Incorporated in 1959).[40] The Nashville-based shoe manufacturer and retailer used the building for its shipping and warehouse department until 1985. In 1987, Gene McLain,

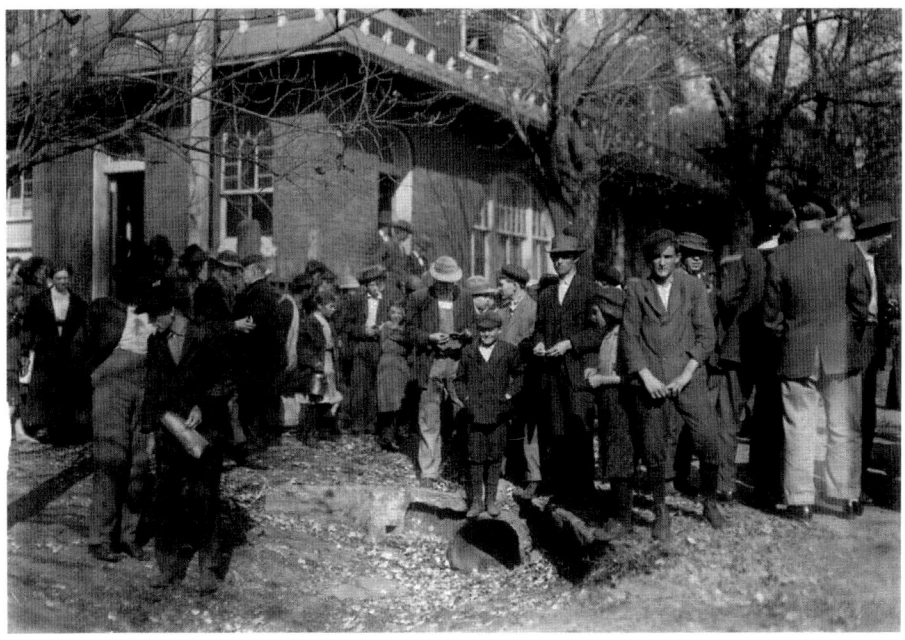

Closing hour, Saturday noon, at Dallas Mill, 1910. *Library of Congress*.

a real estate broker in Huntsville, purchased the property and used the buildings as warehouses.[41]

Legend has it that the ghost of a former millworker has haunted the building since the 1920s. The man was burned and suffocated by still-burning cinders while cleaning out a boiler. Some workers reported hearing noises and seeing shadows in the dark, cavernous boiler room.[42] Oddly enough, the poor man's demise would parallel that of the mill's destruction years later.

THE VILLAGE

As was true with all four of the predominant mills in Huntsville, it was the mill village and village families that determined the financial prosperity of the enterprise. Local businessman Oscar Goldsmith, treasurer of Dallas Mill until his death in 1937, was a major stockholder. He also served as president of the Huntsville Land Company, which was instrumental in the development of East Huntsville. The purpose of the company was to purchase land and construct housing for millworkers. Under Goldsmith's direction, the company began to build what was then known as Lawrence village, consisting of worker housing just east and south of the mill. Families were charged one dollar per room per month for rent.[43] Stores, churches, a post office, a school and a blacksmith shop were also built. Two nurses were employed to attend to the health of the workers and their families. The relationship between the mill managers and employees was both paternal and dictatorial, one that provided for the health and well-being of its workers as well as discipline and structure, at times bordering on severe.

There is no doubt that wages were low and hours long. When the mill opened, the average pay was around fourteen dollars per month, and employees worked twelve-hour shifts.[44] But there was also a sense of camaraderie that extended beyond the factory walls. Though work was hard, the housing and amenities made for an easier life than what many of them had come from, and not one that they had a thought or desire to leave. As author John G. Van Osdell Jr. illustrates in his dissertation, "Cotton Mills, Labor, and the Southern Mind: 1880–1930":

> *It was theoretically possible that a man's mother might attend a pre-natal clinic established by the mill, that he be born in a mill-owned hospital and delivered by a mill-paid doctor, that he be educated in a mill supported*

school, married in a mill subsidized church to a girl he had met in the mill, live all his life in a house belonging to the mill, and when he died be buried in a coffin supplied, at cost by the mill in a mill-owned cemetery.[45]

The Dallas Mill village eventually grew to nearly four hundred houses and seventy-four tenements. It extended south from Oakwood Avenue to O'Shaughnessy Avenue and west to Dallas Avenue. The first homes had no indoor plumbing or electricity. Before indoor plumbing, bathwater was warmed on the coal stove. In 1912, 120 indoor toilets were installed—unfortunately they were all connected, so when one person flushed, every toilet in the village flushed. In 1921, homes became lit by electricity.

Many families had a small garden and even a cow that they could take to the pasture on the other side of Oakwood Road to graze. In 1907, there was a city ordinance against having cows on the streets. The mayor at the time gave instructions for the strict enforcement of the ordinance. Cows were confiscated and several of the owners placed under arrest when they appeared to pay the fine for impounding.[46] This incident was just another example of the "townies" against the "lintheads" mindset that existed at the time.

Villagers and city dwellers didn't interact much unless they had to, and since factory workers and their families had almost everything they needed supplied by the mill, leaving the village was seldom necessary. There was a local barbershop, stores, churches, a school and the local YMCA, which served as centers for spiritual, physical, social and educational activities. Villagers sometimes showered at the Y. The Y had a steam room and shower facilities, two bowling lanes and a basketball gymnasium that was said to be one of the best in north Alabama.

Photos of local basketball legend Newt McGuinness and his team members, as well as other Dallas sports teams, adorned the walls of the Y. Eighty-six-year-old Geraldine Walker sat among her scrapbooks filled with news clippings and photos of McGuinness—the father she never knew. Walker's mother, Nola Wright, went to work for Dallas Mill at age thirteen. She married Newt McGuinness at age twenty-two. They were married seven years, had one child and and were expecting another when Newt was shot in a scuffle with Doyle "Soup" Goodson, another Dallas resident and well-known moonshiner. "Daddy didn't press no charges because he said it was an accident," said Walker. McGuinness had been a prominent member of the Dallas basketball team, coached by H.E. "Hub" Myhand, who came to Huntsville in 1927 as the physical director of the Dallas Manufacturing

Company. Coach Myhand and the 1931–32 Dallas basketball team served as pallbearers at McGuinness's funeral, three days after the shooting. Nola McGuinness raised her two daughters with the help of her parents, who lived on Halsey Avenue. She continued to work for Dallas Mill until its closing in 1949. She passed away at age ninety-one and is buried next to her husband in Maple Hill Cemetery.

Looking at a photo of her father standing proudly in his basketball uniform, Geraldine reminisced, "My sister and I would walk three blocks to see a movie for a dime at the Y on Saturday nights. Coach Myhand was the ticket collector. There were photos of all the sports teams hung on the walls. We never failed to tell Mr. Myhand which one was our daddy, though of course he already knew."[47]

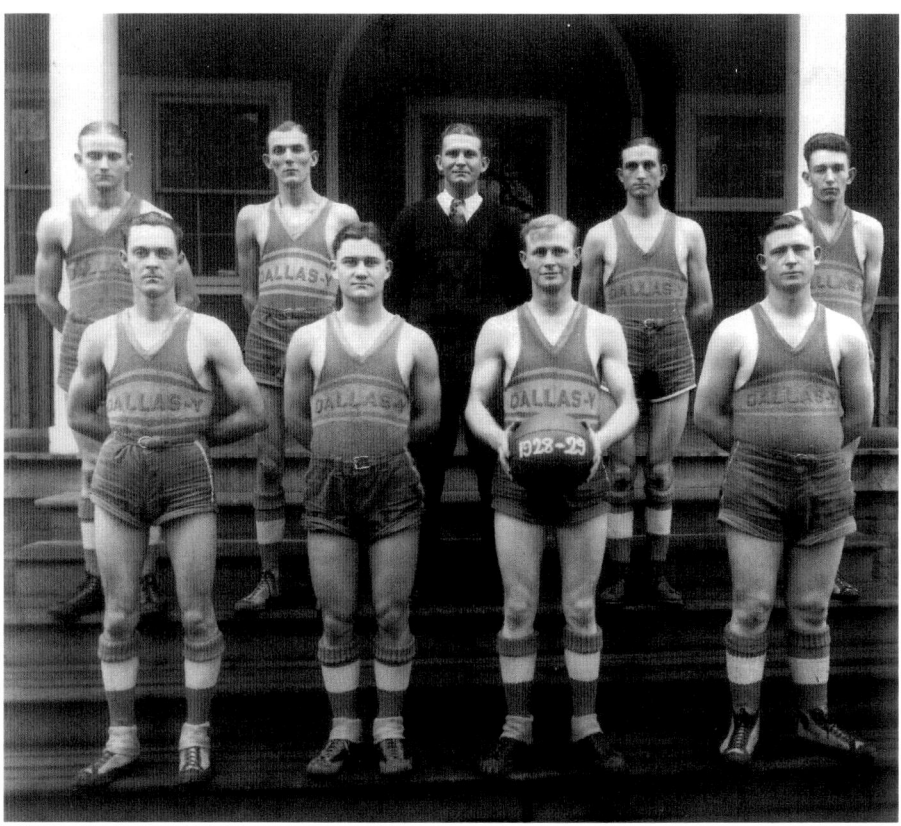

Dallas Mill basketball team, 1928–29. Coach "Hub" Myhand (*top center*); Newt McGuinness to the left of Myhand. *Courtesy of Geraldine Walker.*

The Y went through a number of owners after the mill shut down and was eventually purchased by Fifth Street Baptist Church (now Jackson Way Baptist) as a recreation center. The building was condemned and razed to make room for suburban development in early 1977, after generating over fifty years of memories within its walls.

Dallas Park, built in 1928 on Oakwood Road, served as the baseball field for the semi-pro Dallas Mill teams, also coached by Myhand. "Hub and those that we could get involved in sports as coaches were the babysitters for the village," said Geraldine Walker's cousin Billy D. Harbin. "Hub Myhand was my hero. He was the Bear Bryant of Dallas Village and East Huntsville." The baseball games featuring local mill teams drew loyal crowds of up to six thousand fans from all over north Alabama. It was said that mill managers sometimes hired men for their batting averages. The Huntsville Dr. Peppers, a women's semi-pro softball team, also played at the park from 1937 to 1943. In 1949, the Optimist Club purchased the park. Optimist Park, as it was then called, was one of the few early ballparks open to all races. It was used during the 1950s and '60s for exhibition games by Negro League teams. Jackie Robinson, Willie Mays and other African American baseball legends were frequent visitors to Huntsville.[48]

The School

Children of the mill villages were not allowed to attend junior or senior high in the city without paying nonresident fees, which their families could not afford. Many children went straight from grade school into the mills as laborers. Rison School was built in 1921 to provide quality continuing education to the millworkers' children. The school was named after the mill's general manager, Archie L. Rison. The first principal was P.R. Ivy. Cecil V. "Professor" Fain came to Huntsville in 1915 and served as an instructor and administrator in the public schools of Huntsville and Madison County for half a century. He became the second principal of Rison School in 1926, a position that he held for thirty-two years.

Though not the first mill village high school, Rison was the first at many things. It was the first public school in the area to have a course in Bible study—the course was taught by the ministers of the three local churches: Church of Christ, Baptist and Methodist. It was also, according to Fain, the first public school to have a course in speech and drama, even before it was

Huntsville Textile Mills & Villages

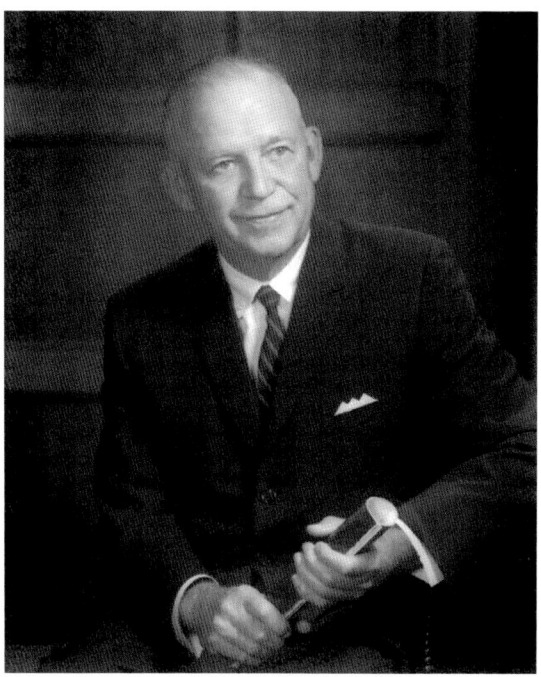

Left: Cecil V. Fain, principal of Rison School, Dallas Mill village. *Rison-Dallas Community website: http://www.rison-dallas.com/principals.html.*

Below: Rison School. *Huntsville–Madison County Public Library Archives.*

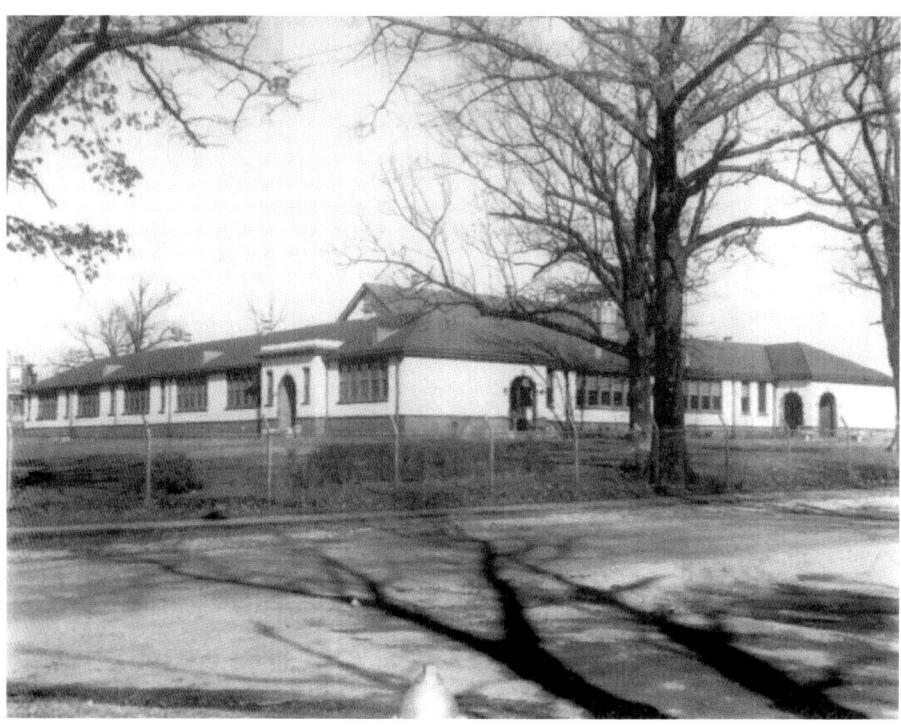

listed in the approved courses by the State Department of Education. This relatively small school, in comparison to some of the larger city schools, had a broad curriculum, including foreign languages, arts, physical education, home economics and industrial arts. Fain was beloved by his students, co-workers and the community. "He didn't see us as the underfed, poor, uneducated mill village children that we were," said one of his former students, Elmore Scoop Hudgins. "He saw us as what we could become—and showed us how to achieve it."[49] *Huntsville Times* columnist Bill Easterling said this of Fain:

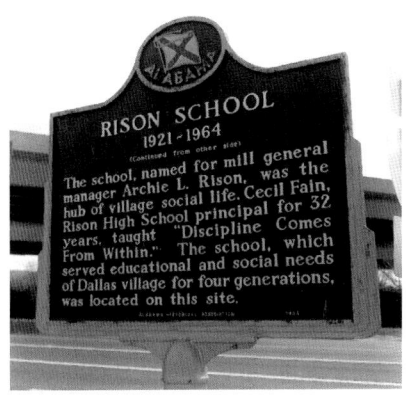

Historical marker at the site of the old Rison School. *Author photo.*

> *He didn't do much: started the first student safety patrol, was principal of eight schools, coached every sport, brought tennis to Huntsville, taught Sunday school for 60 years, organized the county's first PTA group, first Boy Scout troop, first American Legion post...and taught God knows how many boys and girls how to be men and women.*[50]

When necessary, that instruction in maturity took the form of corporal punishment. For instance, when one of the school's star football players was sent to Fain's office for smoking, the principal took the "switch" to the offender:

> *After I hit him two or three licks...I thought it was dust coming out of his trousers....All of the sudden he started jumping up and slapping his derrière. He had a big batch in his right hip pocket of old-fashioned kitchen matches, pretty big matches, and somehow or another that switch had flicked one of those alight and caught all of them alight. It was about to burn him up, so I had to throw my switch and help put out the fire. I believe it's the only time it's happened in the world, as far as I know, that a principal whipped the fire out of a boy.*[51]

Cecil V. Fain died on January 24, 1992, at the age of ninety-six. He was succeeded as principal by Alva Strang Simms and Clarence O. Jones. Jones remained in that position until the school's closing in 1967. Located outside the city limits, these schools came under the jurisdiction of the Madison County School System. They became city schools after 1955, when the Huntsville city limits were extended. In the 1950s, the school became an elementary and middle school, and the three senior high grades were moved to the new S.R. Butler High School. In 1967, the elementary and middle school students were dispersed to three different schools: Chapman, Lincoln and Colonial Hill. The old school building then housed various organizations and was in a state of disrepair when it was torn down in the early 1990s to make way for the construction of Interstate 565.

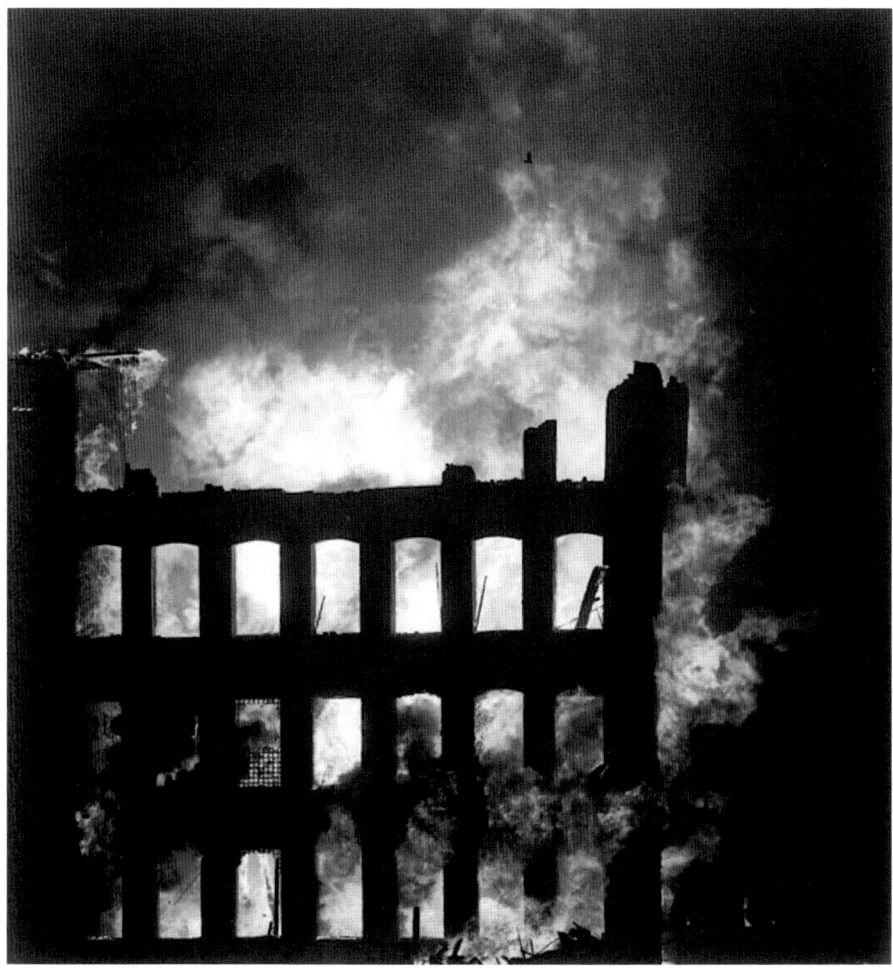

Dallas Mill fire, 1991. *Huntsville–Madison County Public Library Archives.*

The Fire

If the village was the heart of the mill families' lives, the mill itself was the body—a body none thought could be brought down. "It was in good shape when we came in 1954," said former Genesco manager Floyd Drake of the Dallas Mill structure. "Look at those outside walls. They must be three feet thick, all brick. This place could stand up to a tornado."[52]

Fortunately, a tornado never took down the solid structure; it was a fire on July 24, 1991, that finally triumphed over the one-hundred-year-old building. "It looks like the ruins of Rome with the columns left standing there," said Gene McLain, who owned the building at the time of the fire and hoped to restore it. Firefighters fought in vain to save the mill. It took three days for the massive blaze to die out. The battle led to at least one near-death experience. Firefighter Wayne Bridges fell through the first and second floors of the mill into a hole beneath the lower floor. Firefighters had to lower a hose to hoist him out. Law enforcement sources told the *Huntsville Times* that a state prison parolee charged with setting fire to two Huntsville homes in 1990 claimed to have set the mill fire with gasoline, though that was never confirmed.[53] Though many in the community were distraught over the loss of the historic building, including many former workers, one hopes that, at least, the ghost of Dallas Mill was finally at rest.

4
LINCOLN MILL

The Factory

Lincoln Mill had its beginnings in December 1900, as the Madison Spinning Company, located on Oakwood Avenue. Trevanion Dallas, founder of the adjacent Dallas Mill, was a primary investor. The first mill housing and a small school were also added at this time. But by 1903, the locally funded mill required more capital to continue, and outside investors were brought in, including William Lincoln Barrell, a textile baron from Lincoln, Massachusetts. Still strapped for cash, the mill closed just three years later. It reopened in 1908 under the name Abingdon Mills, in honor of Huntsville's first entrepreneur, James White, known as the "Salt King" of Abingdon, Virginia, for his salt-production operation in that city. White had made Huntsville his second home. His business success in the city and the Tennessee Valley made him one of the wealthiest men in America. White personally owned and operated fifty-five mercantile retail stores, mostly along the Tennessee River in Huntsville and other river towns. His business enterprise was the "Walmart" of the early 1800s.[54]

    Despite what appeared to be successful expansions in the early years of production, the mill went bankrupt in 1918. Barrell bought the company at auction later that year, changing the name to Lincoln Mills. The mill specialized in army single- and double-filled duck cotton fabric. A new $2 million, 750,000-square-foot mill building and a one-story dye house were built in 1923–24. They added additional housing and apartments to serve the expanding workforce. Another expansion in 1927 brought

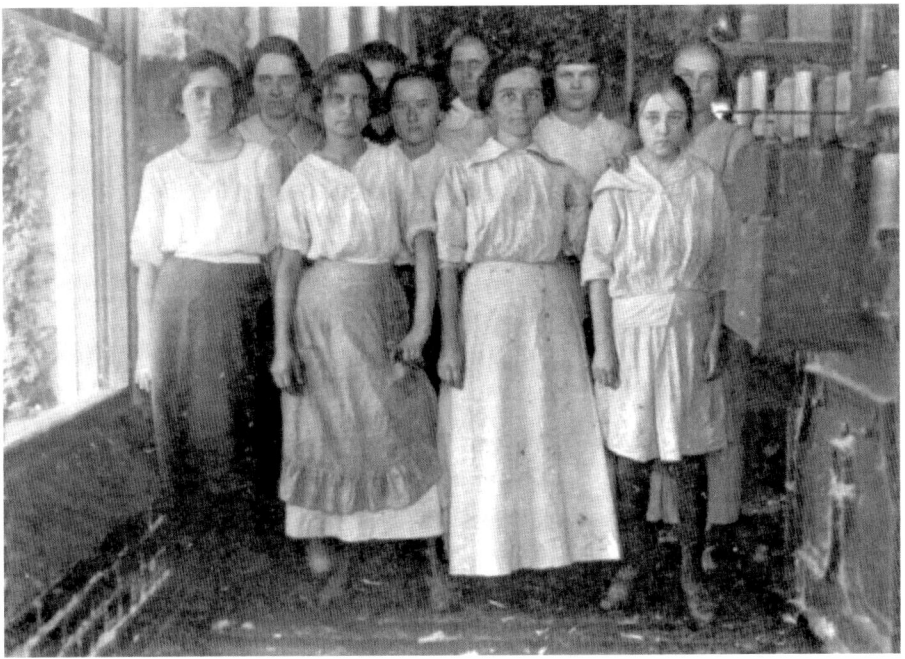

Some of the first spinners at Lincoln Mill. *Courtesy of Larry Lyons.*

the operation to its highest capacity of 120,000 spindles, 1,200 looms and 2,000 employees. The mill shifted completely from steam power to electricity in 1928. In that same year, a mill store and community center were constructed, followed the next year by a larger school.[55] Barrell brought stability and success to the company, and Lincoln Mills became the largest of the city's four textile plants, with five buildings and about 800,000 square feet of production space.

Production slowed during the Great Depression; nevertheless, Huntsville textile mills weathered the Depression better than most other industries by cutting workers' pay, lengthening hours and taking advantage of increasing unemployment to hire men, women and children willing to work for very low wages. In time, this would lead to the great textile worker strike of 1934.

Phillip W. Peeler of Huntsville, with the mill since its beginning, acted as superintendent from 1935 to 1953. After World War II, under Peeler's management, Lincoln Mill won the army's highest industrial award, the "E" award for production above and beyond the call of duty. Peeler was succeeded as superintendent by John W. O'Neal but remained as secretary and director of the corporation.

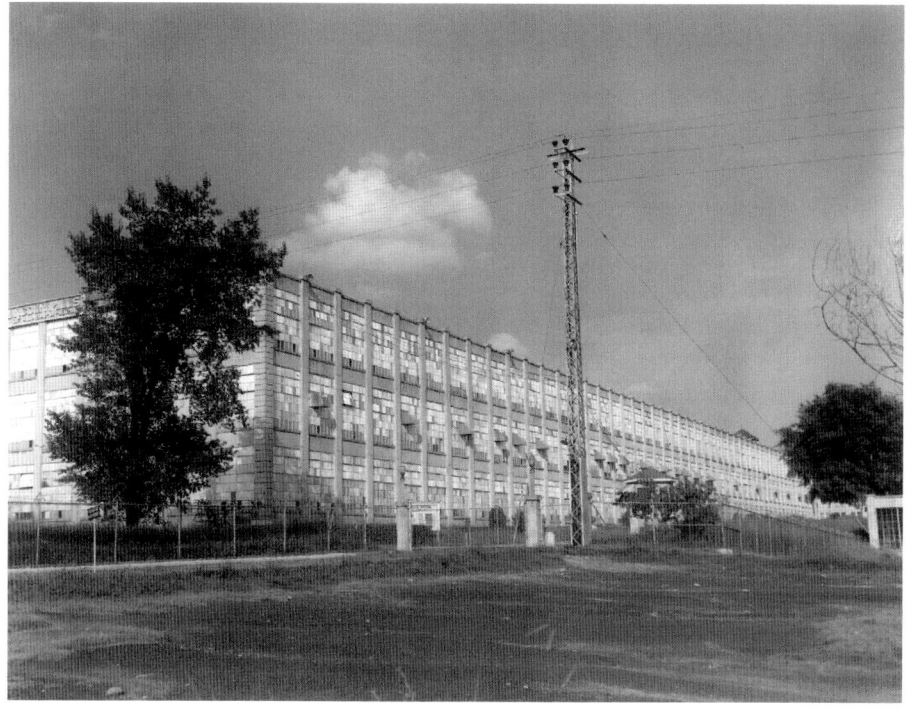

Lincoln Mill, circa 1955. *Huntsville–Madison County Public Library Archives.*

After Peeler's retirement in 1953, a tougher breed of management took over Lincoln's operations. By 1954, the union had filed ten grievances, ranging from increased workload complaints to insufficient bathroom and break time. In 1955, there was a vehement dispute between the local Textile Workers of America and the mill management. The union had a collective bargaining agreement with Lincoln Mills, which provided that arbitration would be used to resolve disputes regarding their relationship. Grievances arose and were processed through various steps, but the union's demands were finally denied by management. When the union requested the agreed-upon arbitration, mill management refused. The case went all the way to the U.S. Supreme Court. The court upheld the union's right to organize; however, the case was deemed moot when mill ownership, as threatened, ceased operations and contracted to sell the properties.[56]

> *The five-month strike at Lincoln Mills has come to an end. With few exceptions, it has been a fruitless walkout that cost its workers more than*

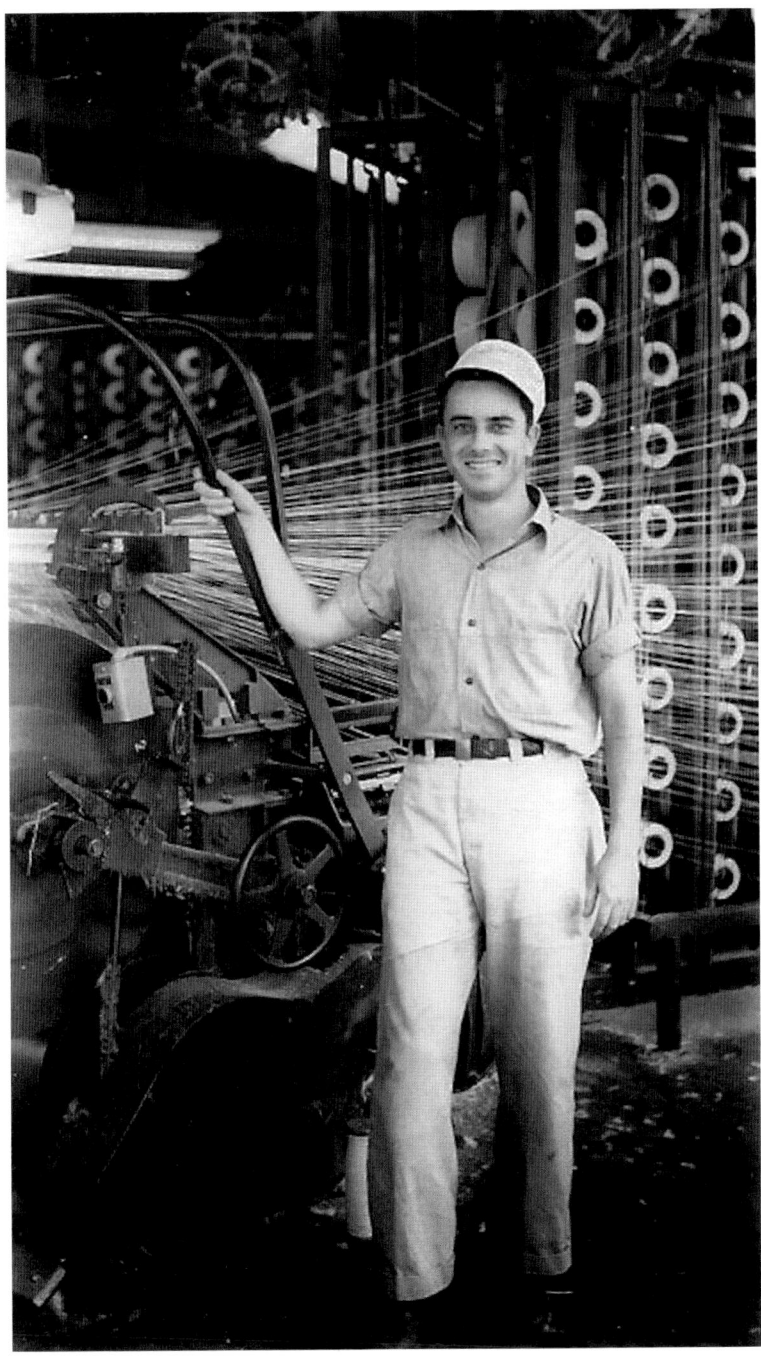

Robert W. Kennedy working as a spool bobber on the threading machine, Lincoln Mill, 1947. *Courtesy of Phillip Kennedy.*

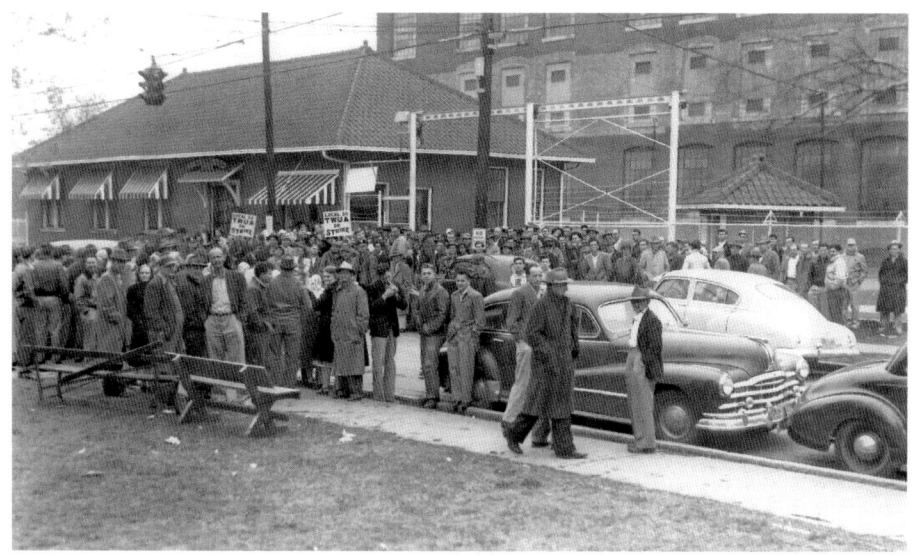

Low wages and poor working conditions were cause for many strikes by unionized textile workers, like this one at Huntsville Manufacturing Company on April 1, 1955. *Huntsville–Madison County Public Library Archives.*

> *$1,000,000 in wages....Both sides lost; neither gained. Added to this is the fact that hard feelings, bitterness and conflict have caused scars that will take a long time to heal or erase.*[57]

Time did not heal the wounds left behind by the strike of 1955, and later that year, the great Lincoln Mills ended its fifty-four-year history.

In February 1957, Huntsville Industrial Center Associates, an alliance of thirty-five local business and government leaders led by Carl T. Jones, purchased the property, renaming it the Huntsville Industrial Center (HIC). Brown Engineering (now Teledyne Brown Engineering) relocated to the HIC building in 1958. The firm performed some of the early contract missile work at the center. The larger facilities enabled the company to continue supporting the army and increase its participation in the nation's developing space program. Brown Engineering relocated to Cummings Research Park in the early 1960s, located on Huntsville's western edge in the cotton fields that had formerly supplied cotton to Lincoln Mills.[58] Milton Cummings, president of Brown Engineering at that time, had grown up in Lincoln Mill village. Cummings was influential in two of Huntsville's most notable industries: space and cotton. As a young man, he had been a successful cotton broker.[59]

The Village

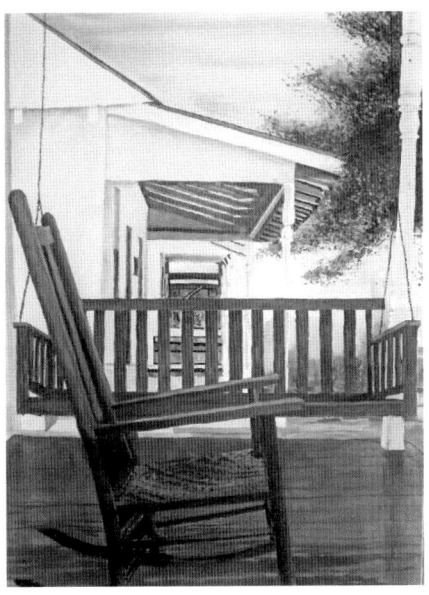

Lincoln Mill village. The front porch was a place where families and neighbors spent time visiting. *Painting by Arni Anderson.*

"The mill built the school, owned the houses, collected the monthly bills for electricity and water, and even passed out the toilet paper," said Margaret Adcock Drake, whose father was a union organizer at Lincoln Mills.[60] But the mill villagers were as much a part of one another's lives as the mill management was of theirs. Most of the residents lived in shotgun-style row housing with a clear path from front door to back door. There was a "cut alley" between the houses so kids could get from one side of the village to the other without going around the entire block.[61] Or, as one-time Lincoln villager Hoppy Boyett remembers, you could just cut through your neighbors' houses until you got home. "Along the way, everybody's mother had supper on the table, and you were more than welcome to grab a piece of this or that as you passed through the kitchen," he said.[62] Arni Anderson remembers spending pleasant evenings on the front porch.

> *When neighbors came over you sat on the porch, visited, gossiped, spit, chewed and generally people-watched and talked about everybody not present. They weren't nosy, they were just interested in what was going on with their neighbors....I recall Monopoly being played for hours on the front porch.*[63]

While almost every home had a porch for visiting on, other things taken for granted in today's homes were missing—such as showers—as J. Curtis Lovvorn, who grew up in Lincoln Mill village in the 1950s, reminisced:

> *About an eighth of a mile from Lincoln Mills, the mill company built a large building, now named the Lincoln Center. The building housed five*

businesses. The Lincoln houses didn't have bathtubs or showers in them, but the barbershop had 12 to 15 showers in the back of it. Mr. Ollie Johnson operated the barbershop. Ten cents for a bar of soap, a bath towel and plenty of hot water, would get you a bath and a locker to put your clothes in while you were taking a shower. So for 10 cents for a shoeshine and 25 cents for a haircut, all total 45 cents, a man could go to the barbershop and go away looking and smelling like a new man.[64]

Lincoln Mill villagers shared their homes, food and sometimes even a shower; they accepted one another as family but were not as well received by their middle-class neighbors in the city. The communities were associated with poverty and crime, and the villagers were scorned as lintheads. The Huntsville Railway Company's electric streetcars aimed to connect the villagers to the city. From 1901 to 1931, Huntsville had two streetcars that connected Dallas Mill, Five Points, downtown and Merrimack, which villagers often rode into town on the weekends to do some shopping. If the merchants disparaged them, they did not turn their noses up to a little of the millworkers' hard-earned money.[65]

One would think that being ostracized by haughty city folk would have strengthened the relationships among the residents of the various mill villages, but this was not so. There existed sometimes friendly and sometimes fierce rivalries between the villages, in particular between the young people of Dallas and Lincoln Mills, who were zealously protective of their own

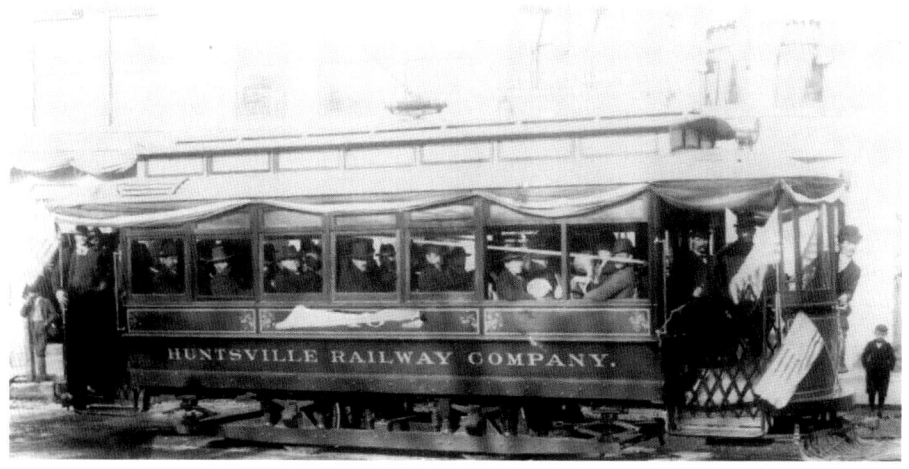

The first streetcar in Huntsville, 1900. *Courtesy of Larry Lyons.*

turfs. The railroad culvert at the north end of Cottage Street separated the two villages. The creek ran on both sides of the railroad tracks dividing the villages. Lincoln and Rison School students regularly had "battles" over the tracks, usually with rocks or BB guns as ammunition.[66]

The rivalry between the two villages had a healthier outlet within their competitive sports teams. Each of the area mills had a recreational baseball team that would play against one another. "All of the teams were pretty even and the competition was pretty lively," said Dallas Mill recreation director Hub Myhand. "I never did see it get out of hand, but we had some pretty good scraps. And everybody in town seemed to be pretty involved in it."[67] Eventually, in 1935, Dallas and Lincoln joined forces and formed a semi-pro team of their most talented members, named the Redcaps. A vintage team with that same name still plays today.[68]

The School

Besides keeping youngsters active in sports, keeping them in school and arming them with an education was another way of curtailing mischief. The Lincoln Mill School was completed in 1929. When Phillip Peeler was manager of the mill, he often filmed activities at the school and on the ball field. "There wasn't a kid in that village he couldn't call by name," said one-time Lincoln School athletic director Obie Johnson.[69] The mill managers were very proud of their schools and were sympathetic to the needs of the children of their employees. Edward Anderson, a part-time Baptist minister, was the school's principal from 1930 to 1944 and again from 1944 to 1965. For the eight years in between, he served as superintendent for Madison County Schools. Although Lincoln School was company-owned, it was staffed and administered through the Madison County Board of Education.[70] Anderson was a strict disciplinarian and did everything he could to keep kids in school. He confessed he used to shave at night so he could use his rough stubble the next day to keep unruly boys in line. "Clamping an arm around the neck of each warrior, the man raked his beard across their faces a few times and turned them loose. They gave no more trouble that day." Anderson went above and beyond the call of duty and even took the boys camping in summers to compensate for the Boy Scout activities and other outings their parents could not afford.[71] Peeler, Anderson and Johnson worked together to

Huntsville Textile Mills & Villages

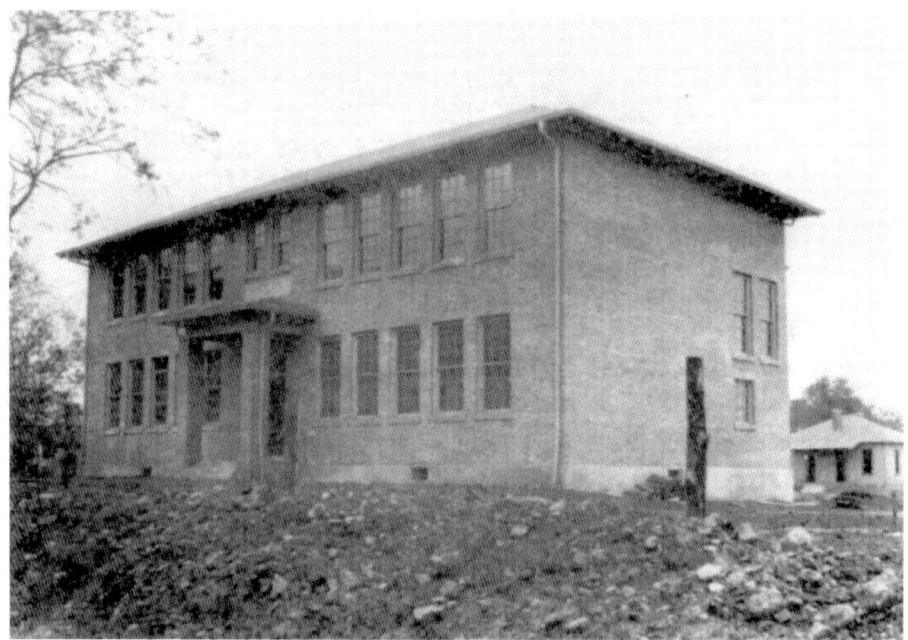

Early photo of Lincoln School. *Courtesy of Larry Lyons.*

make sure the children of Lincoln village were educated and stayed out of trouble. "Mr. Peeler called me in one time and he says, 'Obie, can you help me? I've got a problem. These kids is [*sic*] just tearing the hell out of the village.'" This inspired the two of them to begin the city's first Little League baseball team. Former Madison County commissioner Tillman Hill was on that team. His mother made his ball mitt out of duck cotton produced in the mill.[72]

Anderson went on to build Lincoln into a high school, which became an accredited institution. After his retirement, the school became a junior high. At the time of its closing, the building served as an elementary school for 150 students. It was one of the highest-performing schools in the city. However, the school board decided its maintenance costs were too exorbitant, and the oldest school in Huntsville closed in 2010. The building was sold in 2011, and a private, church-backed school, Lincoln Academy, began operating in the structure the following year.

The Fire

In February 1980, the largest fire in Huntsville's history destroyed much of the eighty-year-old complex but spared the school. Of the five largest fires in Huntsville—the Monte Sano Hotel, Stone Middle School, Dallas Mill, Lincoln Mill and the fairgrounds—the burning of the Lincoln Mill structure was the worst. The whole neighborhood had to be evacuated. Virtually the entire city fire department battled the blaze, which took three days to burn out. "This is the kind of fire movies are made about," said Jay Gates, spokesman for Huntsville Fire and Rescue.[73] Mill building No. 3 and the Dye House, conceived and built as fireproof with a concrete rather than wood beam infrastructure, fulfilled their design intent and survived the fire. The Well House and Chemical Vault were on the southernmost side of the site, and these, along with the Lincoln Mills Headquarters Office, survived as well. These remaining buildings were sold to Robin Ebaugh, one of the sixty tenants left homeless by the fire. The Ebaugh family owned

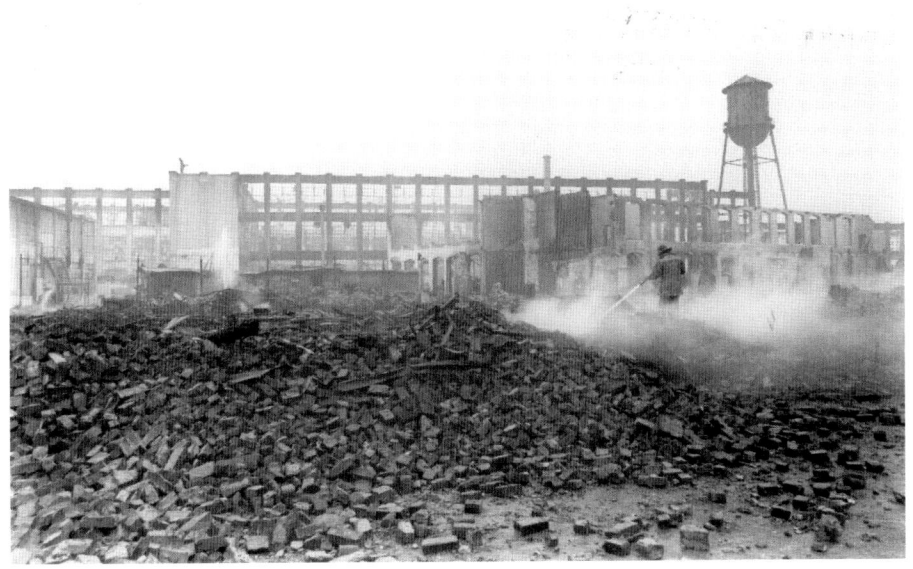

Aftermath of the HIC Building fire, 1980. *Huntsville–Madison County Public Library Archives.*

the buildings from 1982 to 2007, renaming them the Downtown Industrial Complex. While undergoing renovations, the building space was rented for storage and small offices. Huntsville ophthalmologist James Bryne purchased the building in 2007 and began extensive interior renovations as an office complex, honoring the building's beginnings as a textile mill with the name Lincoln Mill Office Campus. The cause of the old Lincoln Mill building fire is still unknown.[74]

5

MERRIMACK MILL

The Factory

"When it comes to down right hustling, Tracy W. Pratt has few equals," stated an 1899 write-up in the *Weekly Mercury*. "A dozen such men as Tracy Pratt would make Huntsville as big as Chicago in a dozen years."[75] Well, there would be no more Tracy Pratts and Huntsville would never grow as big as Chicago, but there is no doubt that Pratt made it his life's mission to advance the city. The native South Dakotan earned the title "Huntsville's First Citizen." Though it was generally agreed upon that he was responsible for more of the major industries locating to Huntsville than any other man who ever resided here, an act of nature almost kept one of its largest industries—Merrimack Mill—from ever being constructed.

Within one mile of the West Huntsville Cotton Mill, owned by Pratt, were two large plantations, a half mile from Brahan Spring. Pratt thought the location ideal for another cotton mill. He'd heard that the Merrimack Manufacturing Company of Lowell, Massachusetts, was interested in locating a plant in the South and wasted no time contacting the owners to petition on behalf of Huntsville. The Merrimack owners expressed interest; Pratt secured quotes from the landowners, and Merrimack representatives arrived for inspection. But a torrential rainstorm had left behind flooding like never seen before, leaving Pratt's perfect site a bog. The irritated Merrimack men are said to have asked Pratt, "Did you think we wanted to locate our plant in the middle of a lake?"[76] They left, disgusted. Undeterred, Pratt secured letters from the city's mayor, an Episcopal priest and a local judge, certifying that the flood was a freak occurrence. He then got on a train and hand-delivered the letters to Boston. His tenacity paid off, the representatives revisited the site on a beautiful spring day and they were persuaded to make it the location of their southern mill, supported entirely from outside investors.

Citizens celebrated our country's independence and groundbreakings for both Merrimack Mill and the new streetcar line (for which Pratt invested virtually all of the necessary capital) on July 4, 1899. Some experienced millworkers from Georgia and South Carolina came to work at Merrimack, but most were farmers and their families from Tennessee. The steam-powered factory began operations on July 9, 1900, with 750 employees, 25,000 spindles and 1,800 looms.

> *Workers in the mill included white men who did the heaviest and best paid work and served as engine and pump room operators, card fixers and spindle plumbers; white women who worked as spinners, loom operators, rope layers in the spinning room and weavers; black men who did menial jobs such as sweeping, cleaning up and unloading cotton bales, and who were not permitted to live in the village; and white children as young as seven whose alternative, at least to one observer, was a life of "sloth and degradation."*[77]

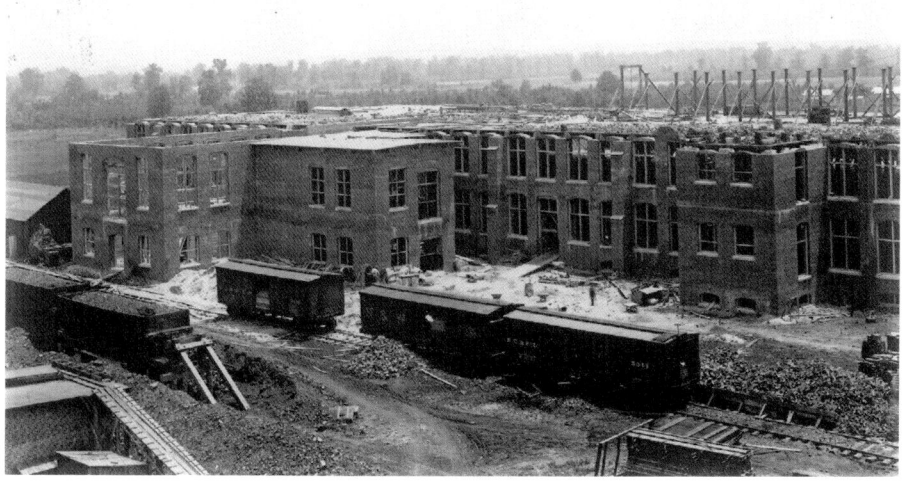

Merrimack Mill building no. 2 under construction, June 18, 1902. *Huntsville–Madison County Public Library Archives.*

Wages of three to four dollars a week were paid to workers in gold coin. The workers presented metal discs to the company's paymaster inscribed with their identification numbers in order to receive pay.[78]

With the addition of a second mill in 1903, the factory became one of the largest in the country—with 90,000 spindles and 2,900 looms—and was dubbed the "Monster Mill" by the *Weekly Mercury*. In 1905, one of Huntsville's most famous and beloved citizens came to the city to serve as agent for Merrimack, relieving George T. Marsh. Philadelphian Joseph J. "Big Joe" Bradley arrived from Georgia, where he was a mill superintendent. The business prospered under Bradley. A 1916 report showed production of approximately one million yards per week of print cloth, percales, organdies and khaki. The mill continued to thrive through the 1920s, when it converted to electricity. Big Joe remained general manager until his death in 1922, when he was succeeded by his son Joe Bradley Jr.; following him was Henry McKelvie, who was succeeded by A.D. Elliot, who came in 1945.

As it did at the other mills, the 1930s brought times of challenge, unrest and despair. Workers were living on as little as three dollars a week during the Great Depression. As they became disenchanted, union membership went up and strikes grew more common. Merrimack shut down in late 1937

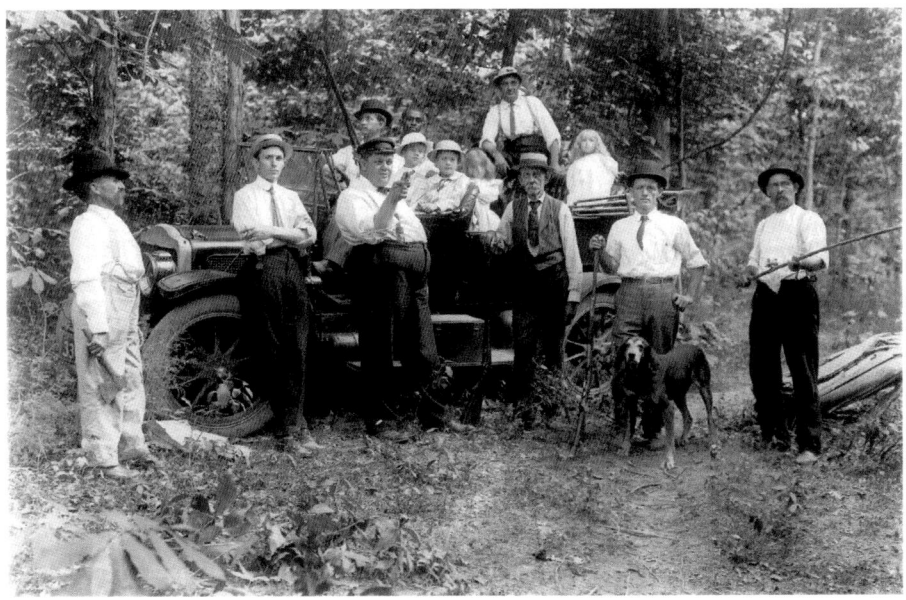

Staged hunting scene, year unknown. "Big Joe" Bradley Sr. (*center*) is pictured here holding a pistol. *Courtesy of June Golden.*

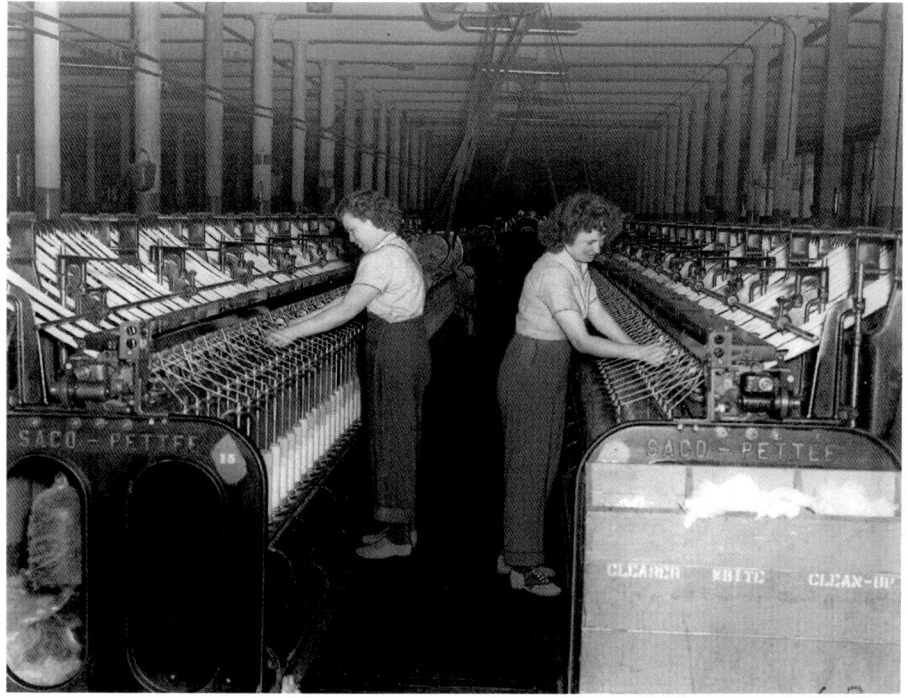

Spinners at Huntsville Manufacturing, 1940s. *Huntsville–Madison County Public Library Archives.*

and remained closed for all of 1938. "We went to Mobile and lived for four years because daddy was so upset about what was going on at the mill," said Doris Holmes Brown. "After the strike, when he heard the mill had opened up again, we came back up here in 1939. He interviewed and because he'd been assistant to the man who was overseer, they gave him that job."

World War II ignited a resurgence in business during the 1940s, when the company began producing uniforms and other fabrics for the war effort. After the war, Merrimack's peacetime production dropped to a low. The plant sold to M. Lowenstein & Company of New York on November 25, 1945.

January 13, 1946, was the last day of operation for Merrimack Mill. The following day, under a banner that read "Welcome to the Huntsville Manufacturing Company," employees returned to work. Woodrow E. Dunn served as vice president and general manager. Lowenstein & Company spent millions of dollars in renovations, doubled the workforce and increased production—the mill was again a major competitor in the industry. In

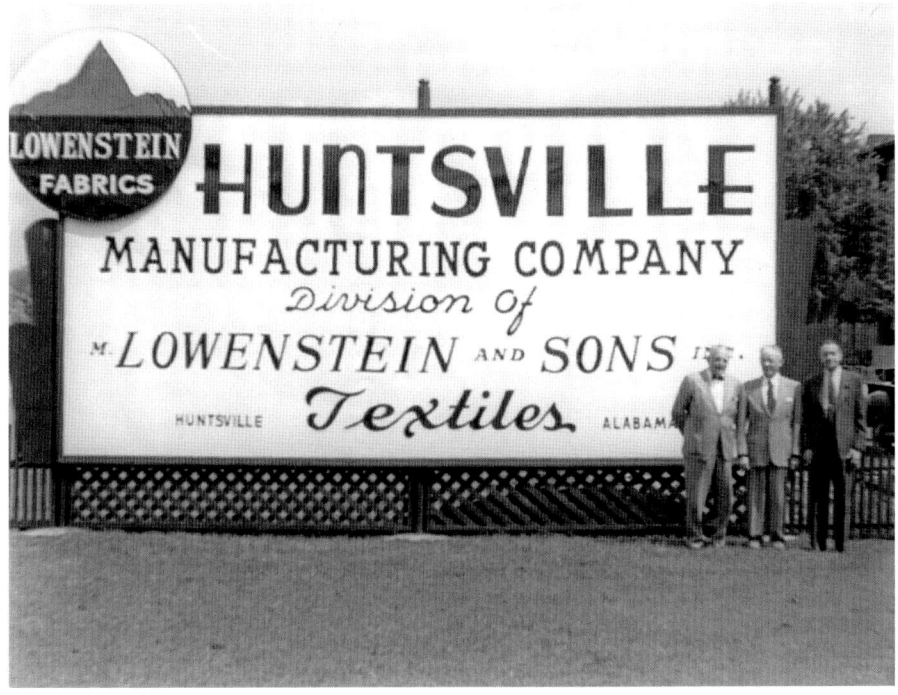

Leon Lowenstein, A.D. Elliot and J.J. Lyons at the opening of Huntsville Manufacturing. *Huntsville–Madison County Public Library Archives.*

September 1947, the mill was judged by the Financial World Annual Report Survey to be the "Best in the Textile Industry" and won an "Oscar of Industry" trophy.[79] In the 1950s, equipment was overhauled, along with other remodeling and improvements, including air conditioning in the weave rooms. In 1955, production was at its peak, with 145,896 spindles, 3,437 looms and 1,600 employees.

THE VILLAGE

In 1900, at the same time as the first mill building was being erected, the village's first houses were under construction. The homes at Merrimack were more spacious than many other village homes. The houses were either duplexes or apartments with large front porches. In the early days, the rooms were lit by oil lamps, and water was piped in from Brahan Spring to a hydrant at the rear of each lot. Homes built before 1937 had concrete privies out back, where the coal was also stored. As it was in other mill villages, many residents tended small gardens and owned a few cows, chickens or sheep. Pens for hogs were provided a little distance from the village; however, villagers said that when the wind blew a certain direction they seemed closer. Wagons delivering baled hay and straw passed regularly. In the 1920s, the village modernized with a sanitary sewer system, sidewalks, curbs and gutters.[80] In 1922, all the mill homes were wired for electricity, and by 1925, there were 279 homes in the Merrimack Mill village.[81]

From the beginning, Tracy Pratt realized that a plethora of amenities would draw workers and their families to the village, as is indicated in an article in the *Huntsville Republican* dated January 6, 1900:

> As the Merrimack Mill nears completion, the attention of the public is attracted to West Huntsville.... [L]ast week we made notice of the fact that a gentleman had begun the erection of a building to locate a steam bakery. There is [sic] also good prospects of locating a bank, meat market, grocery,

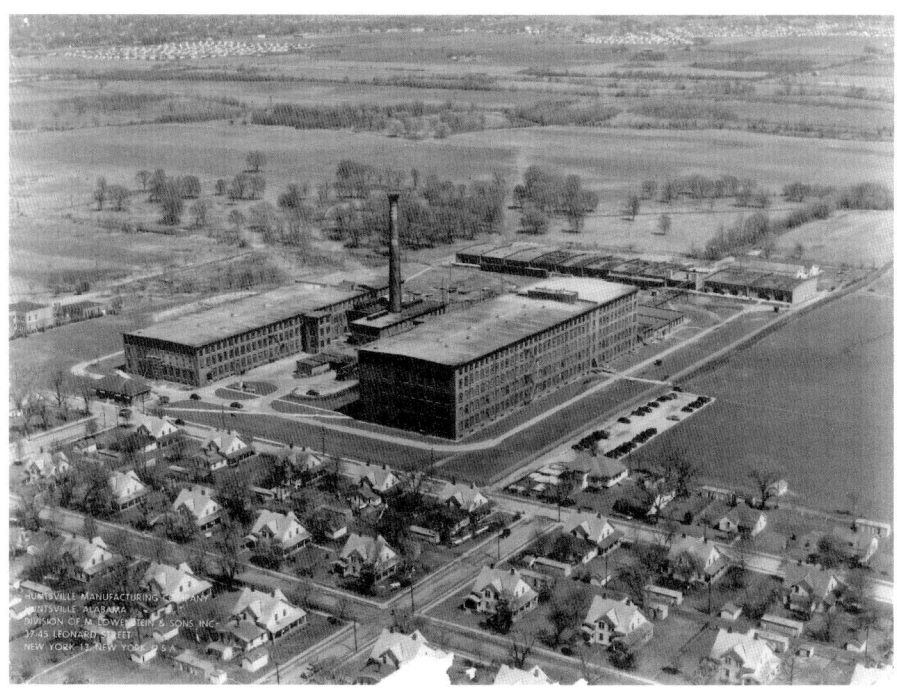

Huntsville Manufacturing Company and mill village housing, circa 1948. *Huntsville–Madison County Public Library Archives.*

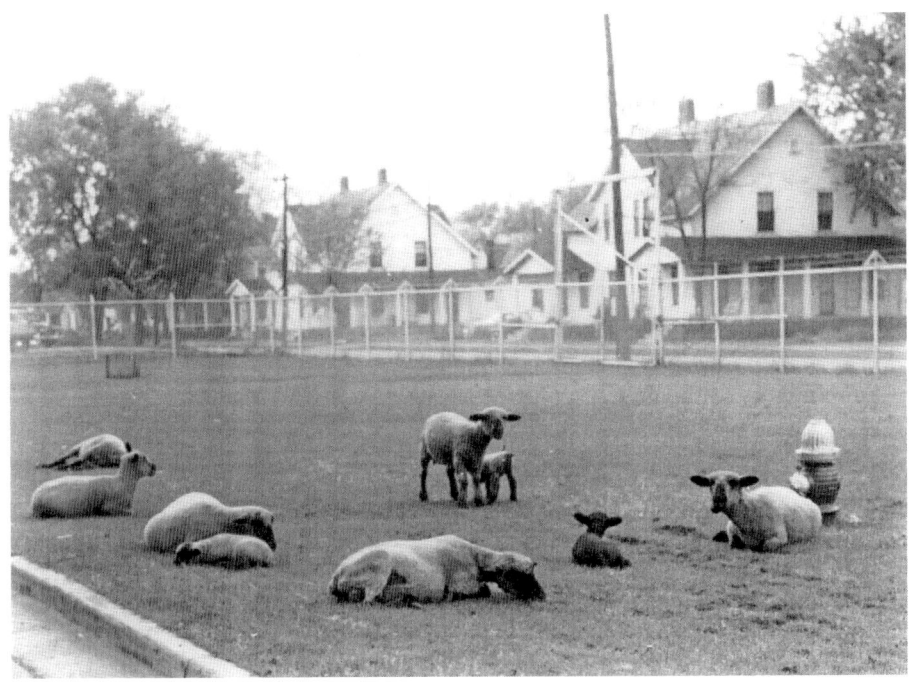

Sheep grazing in the Merrimack pasture. *Huntsville–Madison County Public Library Archives.*

and a general store. All of this shows that our recent prediction that West Huntsville was going to be one of the best suburbs to our city is getting more prominent every day. We would advise investors to keep in touch with West Huntsville.

The first company store opened in 1902, in an existing two-story wood-framed building. In 1913, the mill acquired another building, and the store was greatly expanded in 1920, incorporating the two structures. The twenty-five-thousand-square-foot building housed the community store, a barbershop, a modern café, a gymnasium, a theater and other amenities. The store was run for many years by a man named Searcey McClure, and many residents of the village referred to the store as Searcey's.[82] On payday, even after a long shift at the factory, workers would go home, clean up and head back to the store for dances, socializing, a movie or a basketball game. Eighty-nine-year-old William Deward "Bill" Brown's grandfather, Shorter Brown, helped to construct the mill and later became a gatekeeper. Bill remembers his visits to the mill store. "There was a rock, a cement abutment, across from the store and the men would sit there in a row and ogle and whistle at the girls," he said. His wife, Doris Holmes Brown, added, "My sister said a girl would not dare go over there or she'd get a bad name."[83]

"The mill village system was a benevolent serfdom," said Brown. The workers' lives were controlled both inside and outside of the mill. Corporate principles stressed the necessity of religion to keep sinful natures in check and moral standards high, and managers wasted no time in seeing that their operatives were properly immersed in Christian tenets. In June 1900, the *Republican* announced that a Miss House would serve as missionary to the Merrimack employees. Miss House arrived later that summer and also established a kindergarten and provided other educational mission services for the community. Later, various Protestant denominations would erect centers of worship, including Baptist, Methodist and Church of Christ. Pentecostal churches were discouraged, as they might elicit "emotional outbursts."[84]

Mill manager and village advocate Joe Bradley made sure that the workers and their families were taken care of physically as well as spiritually. Smallpox and malaria outbreaks had become problems and posed a threat to the villagers—and thus to production. The first mill hospital was established at Merrimack Mill village in 1913 in a small cottage on South Broad Place, with one resident doctor and two nurses. By 1916, a full-scale hospital building was complete, and it was even made available to city residents during times when Huntsville Hospital experienced critical overcrowding. The village hospital was equipped with a first-aid room, a recreation room, showers, two hospital rooms, an operating room and a doctor's office. Later, a dental clinic was added.[85]

Eighty-six-year-old Maebelle Winkles grew up in Merrimack Mill village with her eight brothers and sisters. Her mother and father, Katie Lee and John Oliver Marks, both worked at the mill, her mother as a spinner and her father as a loom fixer and a gate watchman.

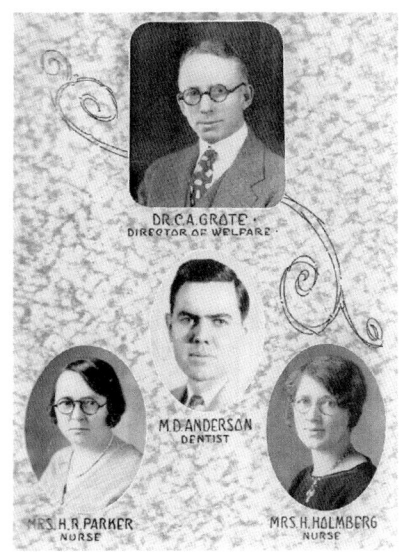

Merrimack Hospital medical staff, 1931. *Huntsville–Madison County Public Library Archives.*

Maebelle's mother started working in the mill at about the age of eight: "Mama worked in what we called the 'little mill.' When they built mill number two they moved her over there....My oldest brother was born in 1912 and when he was born she quit work to raise children. But Daddy worked sixty some years at the mill."

The recent widow still lives in the same house on South Broad Street that she and her husband, Ernest Lee "Bud" Winkles, shared for the last thirty years. "This house was built for the nurses at the [Merrimack] hospital and the one next door was for Dr. Grote," she said. Maebelle also shared some of the day-to-day experiences of the mill family:

> *We raised hogs and killed six or seven of them a year. Everybody saved their scraps for us and the boys would go around and collect them and then go slop the hogs. Daddy and the boys cut up the hogs and salt cured the meat in big boxes. We didn't have any refrigeration....Mama rendered the lard and stuffed sausage into little white bags that she tied off and hung up in the cold house. If we wanted sausage for breakfast all we had to do was go out back and get a bag....Dr. Grote'd tell my mama, "Katie, be sure to save me all those chitlins." We shared what we had.*[86]

Dr. Grote treated serious mill-related injuries as well as the cuts, bruises and broken bones brought on by childhood mishaps, as related by Bill Brown:

> *In the summer time we got what was called "stone bruises." If you walked barefoot hard enough and long enough you'd get an infection in the heel and you'd go up to Dr. Grote and he'd lance it, and let the pus drain out, and stitch or tape it up and you'd have to hop around until it got well....That was commonplace.*

THE SCHOOL

In 1906, Alabama had a state law requiring children to attend school eight weeks a year, with six of those weeks being consecutive. Until the Child Labor Law was placed into effect, children were allowed to work in the mill year-round, reporting for their shifts at the end of their school day. In Merrimack village, prior to 1919, any schooling took place in home dwellings or small buildings. The first of these small schools was Doutheboys Hall School,

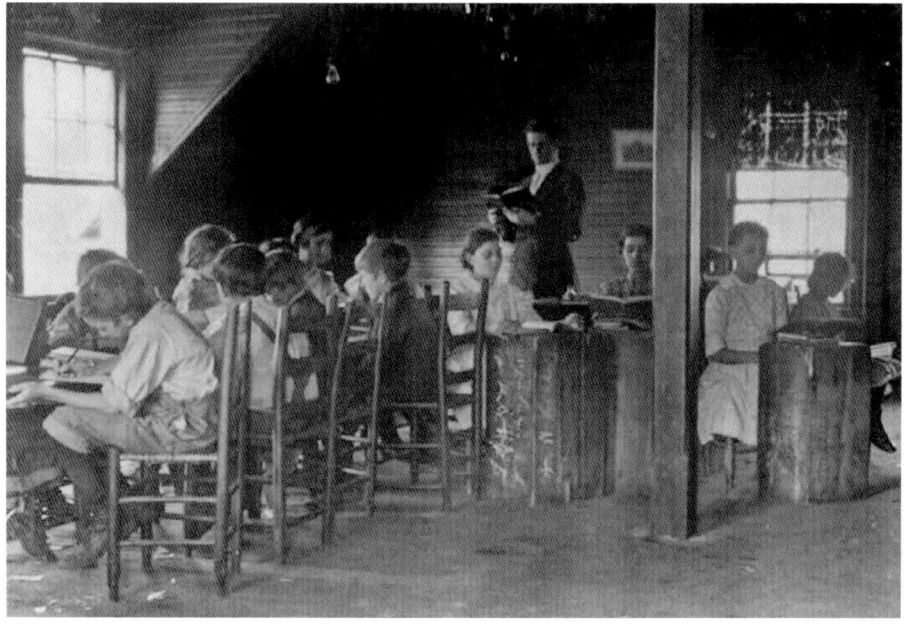

Doutheboys Hall in session, 1913. *Library of Congress.*

constructed in 1913, followed by a four-room school in 1914. In June 1919, a larger building, known as the Joseph Bradley School, was built. The two-story brick structure cost $22,000, had ten rooms and an auditorium and could accommodate seven hundred students.[87]

After returning home from World War I, Cecil Fain agreed to serve as principal for Joe Bradley School. In an interview with the *Huntsville Times* in 1992, Fain recalled the stern words Big Joe said to him upon his hiring. "He said, 'Don't come out here with your head way up in the air, thinking you're better than these people.' I'll always remember that," said Fain.[88] He held the position for four years—also beginning and coaching many of the sports teams—before moving on, later to become principal at Dallas Mills' Rison School. He was succeeded in 1923 by Edward Foyle (E.F.) Dubose. Dubose had attended college in Troy and the University of Alabama and spent a year and a half teaching at Etowah County High School. The story goes that in the spring of 1921, Joe Bradley Sr. was walking past the Etowah County High School and stopped to admire the beautiful school garden. He was so impressed that he wanted whomever was responsible to plant something similar at the Merrimack village school. In September of that year, he hired Dubose. The educator loved watching seeds grow into plants almost as much as he loved watching children mature into young men and women. He served on the faculty of Joe Bradley School first as an agriculture teacher and vice principal before taking over Fain's position.

E.F. Dubose served as the school's principal for forty years, seeing it through its many expansions and renovations. Just three years after Dubose became principal, the school became an accredited institution and went on to become one of the finest consolidated high schools in the Southeast. Dubose was a firm disciplinarian and kept track of what his students did inside and outside of school. One-time resident George Reavis spoke to former *Huntsville Times* columnist Bill Easterling about some of that mischief. "We dismantled a fellow's wagon and then reassembled it on top of somebody else's barn. We swapped a few cows on Halloween. Real mean stuff like that," said Reavis.[89] It's said that Dubose could look a student in the eye and say, "I know what you did and who you are," and they would voluntarily confess their shenanigans.[90]

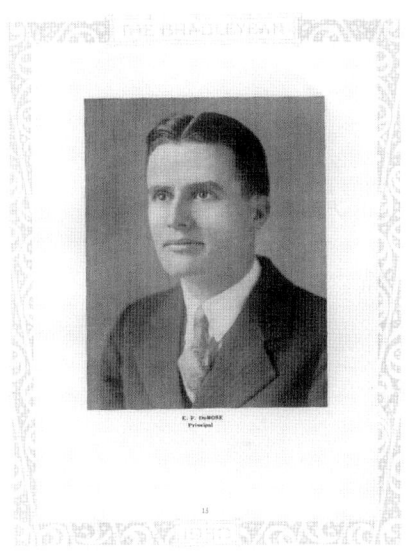

E.F. Dubose, principal of Joe Bradley School. *Huntsville–Madison County Public Library Archives.*

While there might have been mischief, Dubose made sure the children received an exemplary education and hired only the best teachers. He encouraged students to continue their education beyond high school. During World War II, he also rewarded young men who were going to serve their country. Dubose set a decree that any senior who went off to war and was making passing grades in all of his subjects would be awarded a high school diploma when the class graduated. "I was fortunate enough to still be here when we graduated. There were just three boys in the graduation ceremony," said Bill Brown. "Afterwards we immediately went off to war. The others had already gone."[91]

Bill and his elder brother, Clarence O. Brown, went on to attend Auburn University. Clarence received a partial scholarship from the mill while working alternate quarters. He majored in textile engineering, a condition of the scholarship. Clarence was drafted while attending Auburn and completed his degree in engineering at Cornell University after the war. Bill

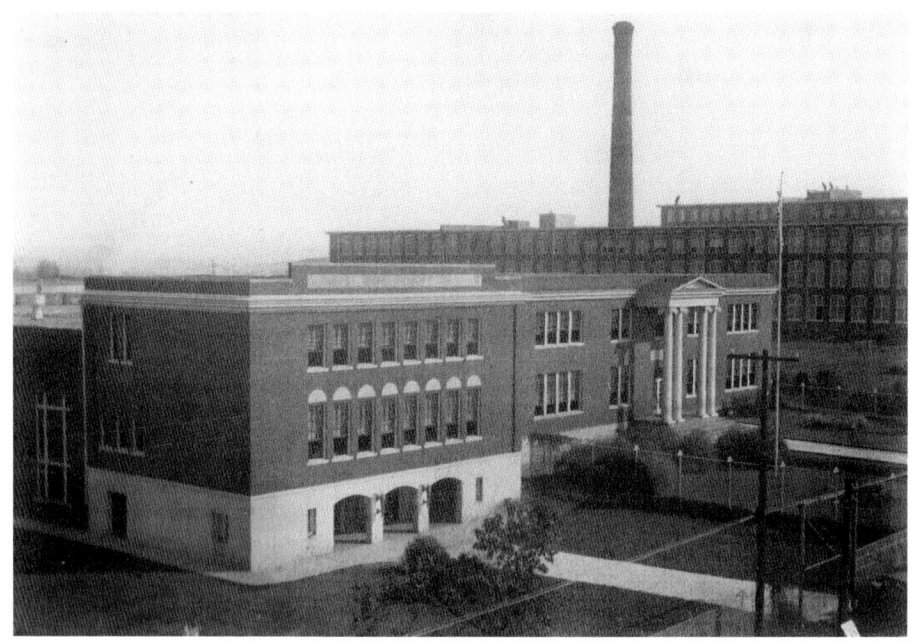

Joe Bradley School, 1930. *Huntsville–Madison County Public Library Archives.*

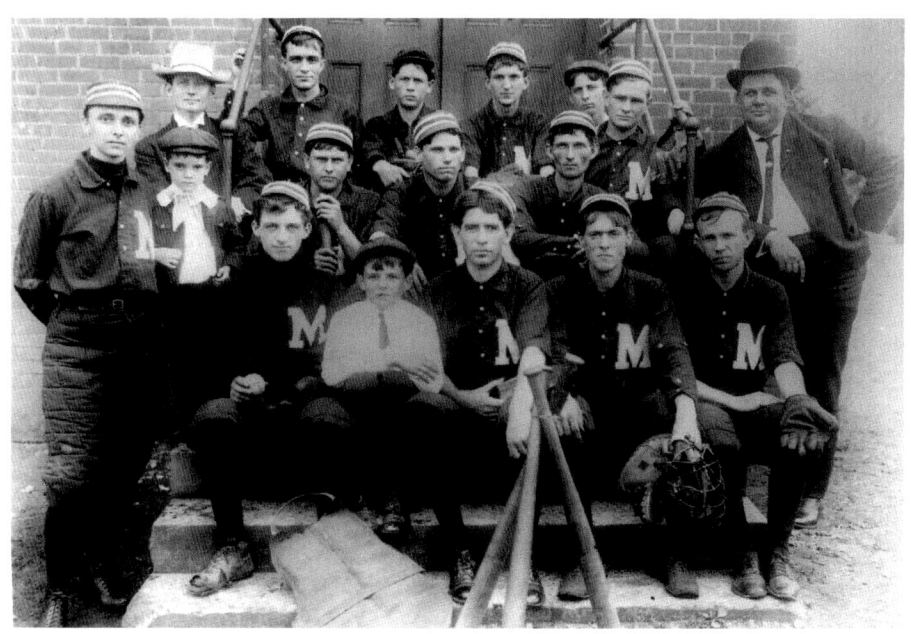

Merrimack baseball team, 1906. *Huntsville–Madison County Public Library Archives.*

received a degree in chemical engineering from Auburn in 1951 and went on to work and retire from NASA. Many of the millworkers' children went on to get college degrees.

As with the other mill village schools, physical prowess was held in as high esteem as mental aptitude. Sports were an integral part of the extracurricular activities and provided entertainment for the millworkers and their families. One of the most recounted sports events was the school's one and only defeat over cross-town rival Huntsville High School in 1947, in front of a crowd of five thousand spectators at Goldsmith Schiffman Field.[92]

In 1951, Huntsville Manufacturing Company deeded Joe Bradley School to the Madison County School System. Beginning in 1952, the school housed first through eighth grades, and high school students were zoned to Butler High School. In 1962, Dubose was moved along with all of Joe Bradley's elementary students and faculty to the newly constructed Ridgecrest Elementary. Dubose served as Ridgecrest's principal until his retirement in 1970, after which he devoted his life to his family and his plants. Joe Bradley School was demolished in 1986. Longtime principal E.F. Dubose died on October 4, 2002, at the ripe age of 102.

HUNTSVILLE'S LAST COTTON MILL CLOSES

Huntsville Manufacturing continued to receive attention beyond the state of Alabama through the 1970s. The upshot of that recognition at times brought unwanted results:

> *Huntsville Manufacturing Company received some flattering but unwanted publicity when an item about it appeared in a Michigan AAA guide several years ago, and tourists began to drop in unannounced for tours. To further complicate matters* Good Housekeeping *mentioned the company by name in an article about interesting things to see in the South.... The many floors of fast-whirring machinery are not conducive to touring children or other curiosity seekers.*[93]

The success of Merrimack Mill and the Huntsville Manufacturing Company has been documented and publicized more than any of the other Huntsville mills, but even with this success, Lowenstein could not circumvent the overall decline of the textile industry as a whole. By the

1980s, foreign imports had increased, and government regulations and restrictions became more stringent. In 1988, Lowenstein sold to Spring Industries of South Carolina, which closed just a year later. On August 26, 1991, after one hundred years of operation, Huntsville's last surviving operational cotton mill was destroyed—not by fire as Dallas and a portion of Lincoln—but torn down and carried off piece by piece by wrecking balls, dozers and dump trucks.[94] One wonders which demise was more difficult for the workers and villagers to watch. "It was sad," said Bill Brown. "Doris and I watched it come down. The smokestack was the last to go. I kept a couple of bricks for souvenirs."

6
LOWE MILL

THE FACTORY

Between 1900 and 1902, almost simultaneously, four mills—the Lowe Manufacturing Company, owned by Arthur H. Lowe of Fitchburg, Massachusetts, president of the New England Manufacturer's Association; the Madison Spinning Company; the Rowe Knitting Company, owned by William H. Rowe and son; and the Eastern Manufacturing Company—arrived in Huntsville. Local newspapers boasted that Eastern Manufacturing Company had the first weaving and dyeing mill in the South.[95]

Lowe Mill, incorporated in 1900, might not have located to Huntsville at all without Tracy Pratt's relentless and ultimately successful negotiations with Arthur Lowe, as evident in this telegram sent to local mill investor J.R. Boyd on August 14, 1900: "Have closed Lowe matter. Will deed at once original terms. Work to commence at once. T.W. Pratt."[96]

Lowe, the fifth mill to be established in Huntsville, began construction at Ninth Avenue and Seminole Drive soon after that telegram was delivered. As with the other mills, capitalization was heavily invested by northerners, with a few local investors, including Pratt, J.R. Boyd and Willard Wellman. Pratt also secured Rowe Knitting Company, located at Ninth Avenue and Tenth Street, that same year. Pratt and Wellman were the only Huntsvillians to serve on Lowe Manufacturing's board of directors. The south section of the mill was built in 1901. In 1903, Lowe Mill absorbed Eastern Manufacturing Company and the following year constructed the north portion, which connected to the earlier portion. Using steam power,

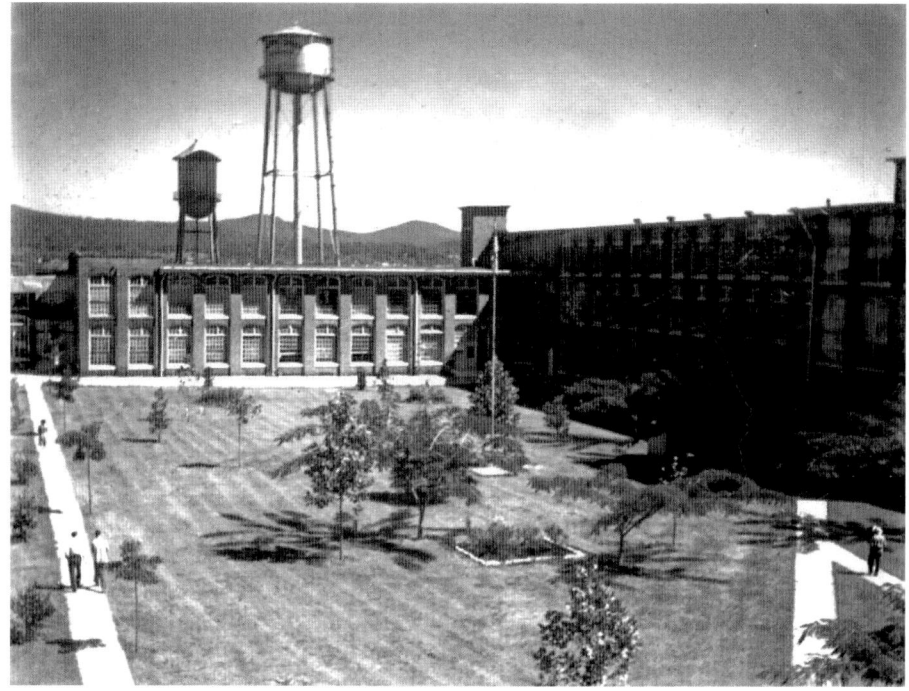

Lowe Mill. *Wikimedia commons.*

the mill produced ginghams, sheeting and romper cloth. By 1906, the mill had twenty-five thousand spindles and by the following year employed nine hundred workers.[97]

In 1909, Arthur Lowe and New Yorker Charles Poor, who had joint interest in Eastern Manufacturing Company, decided to separate their interests. Poor bought out Lowe's interests in Huntsville, and Lowe bought out Poor's interests in the Massachusetts mills and subsequently left the South, though the factory continued as his namesake.[98]

In 1911, Lowe Manufacturing was bought out by the Hunter Manufacturing Company, although it continued to hold on to the Lowe name. By the mid-1920s, the number of employees had dwindled to fewer than six hundred, though several additions, including a warehouse and picker room, were made. Between 1928 and 1932, several of the smaller mills in West Huntsville underwent name and management changes. The debt-ridden Huntsville Knitting Company became Helen Mills, Margaret Spinning Mill became Fletcher Mills and West Huntsville Cotton Mills closed its doors.[99]

LINTHEAD LEGACY

When the Great Depression hit, the number of workers at Lowe Manufacturing declined even more, and the plant began producing solely "grey goods," or unbleached sheeting. The decade brought much upheaval and instability. By December 1932, the mill had declared bankruptcy. Just a month later, the plant started up again under the name Lowe MillsIncorporated, with Donald Comer, head of Birmingham's Avondale Mills, as majority stockholder. On July 17, 1934, the workers of Lowe Mill walked off their shift due to lack of progress to improve working conditions, which would ultimately lead to a national strike of the United Textile Workers of America, to be discussed more fully later. In 1936, the mill changed hands again, with Edwin Green of New York as majority stockholder. The newly named Lowe Corporation dissolved in March 1937 and was sold yet again to Walter Laxton and became a cotton warehouse.

At the end of World War II, in December 1945, Nashville-based General Shoe Company/Genesco Incorporated purchased the factory. During the Vietnam War, Genesco's nearly eight hundred employees produced the majority of combat boots for U.S. soldiers. Genesco continued production until 1978, when Martin Industries purchased the plant, and it served as a warehouse for residential and commercial heating systems. In 1999, the

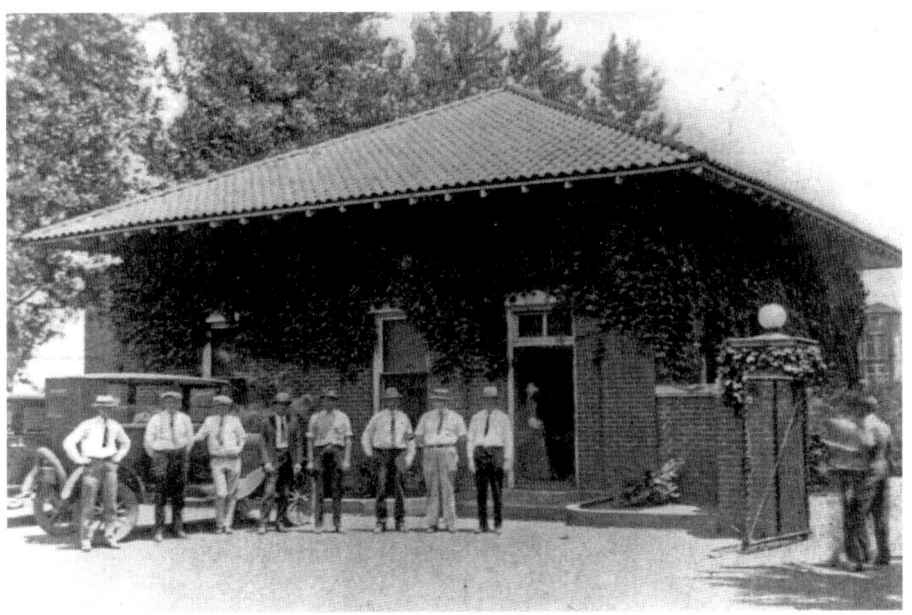

Overseers and office for Lowe Manufacturing, 1925. *Huntsville–Madison County Public Library Archives.*

mill acquired yet another owner, realtor Gene McLain, who kept it for less than two years before selling it to its current owner: Research Genetics founder Jim Hudson.[100]

THE VILLAGE

Residential housing was built concurrently with mill construction. Between the mill and Triana Boulevard Southwest, there would eventually be 349 family homes in an eclectic layout and a wide variety of styles, predominantly one-story frame duplexes, many with pyramidal roofs. Unfortunately, the Colonial Revival–style superintendent's house, which stood at the corner of Seminole Drive for 105 years, was destroyed in 2006.[101] Betty Owens's family lived in a six-room house on First Street and Ninth Avenue. In 1936, her father purchased the home from the mill for a total of $400. "I think he put $65 down and paid $7 a month. We paid the mill $2 a month for our water, and Mother said that was too high."[102]

There is an unconfirmed story that when thousands of unused bombs were returned to Redstone Arsenal after World War II, the tin-lined pine boxes were sold to Lowe Mill residents for twenty-five cents apiece. Because of a lumber shortage, they used the wooden panels of the boxes for floors and roofing. Money made off of the sales is said to have helped to pay to dismantle the bombs.[103]

Residents could have gotten the nails for those floors and roofs at J.C. Brown General Merchandise at the corner of Ninth Avenue and Ninth Street. The store was actually established before the mill, in 1898, by Jess Charles Brown. The original store burned down but was rebuilt in the early 1920s. It was later run by Brown's son, William Albert Brown. Millworkers went to the store to cash paychecks, pick up a few groceries and grab an RC cola, a can of snuff or a pack of cigarettes. It also served as the hardware store and barbershop. One can still read the inscriptions on the outside brick wall of the building where, for decades, boys sharpening their knives carved their initials. The original owner's grandson, William Atwood Brown, remembers hanging out at the store:

> A guy would come in from the first shift at the mill and he'd cough and I'd say "Camel smoker." You could tell by the cough if he smoked Pall Malls or Camels. I'd say, "You're going to kill yourself." He'd say, "I'm going to go one way or another."[104]

Linthead Legacy

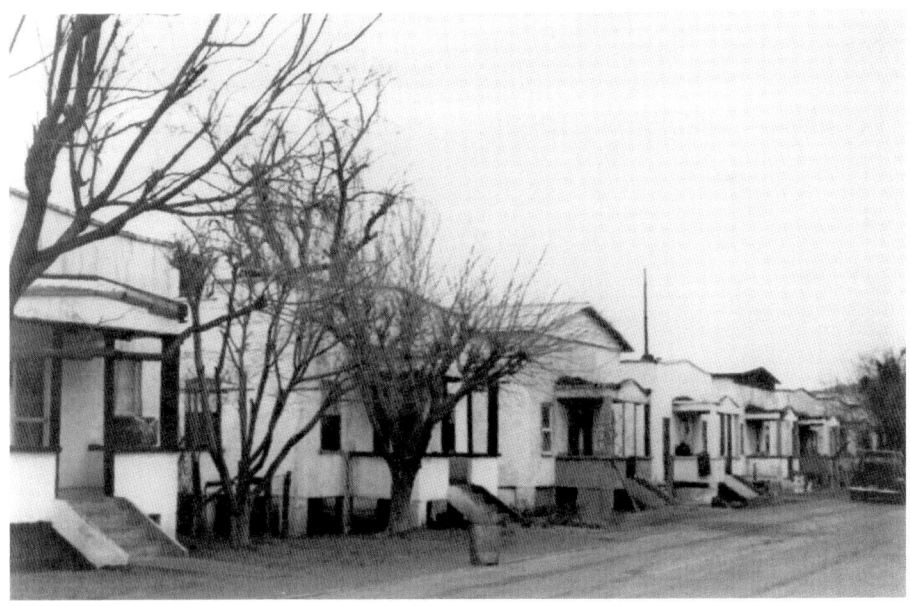

Lowe Mill village homes. *Courtesy of Larry Lyons.*

The store served West Huntsville and the Lowe Mill district until the mid-1990s. In 2015, the entire contents of the store were sold, including fixtures, the original pine flooring and the antique tin ceiling tiles.[105] More than 250 people waited in line to get their piece of Huntsville history.[106]

Just west of J.C. Brown Merchandise was the low-income shantytown known as Boogertown. To the Boogertown kids, mill villagers seemed like rich folks, and it was true that those on the "right side of the tracks" had many privileges and opportunities. There were churches, the Art Deco Centre Theatre, a YMCA and a general store all available to the families of Lowe Mill village and the residents of West Huntsville.[107] The West Huntsville YMCA, housed in the McCormick building, touted itself as "one of the largest and best-equipped industrial young men's Christian association in the South." It was organized in 1915. Mary "Virginia" McCormick, heir to the McCormick harvesting machinery fortune, donated the land and the building. Executives from Lowe Manufacturing Company, the Huntsville Knitting Company and the West Huntsville Cotton Mills served on the YMCA's board of directors. There is no doubt that the YMCAs in the mill communities played a big part in keeping villagers happy by providing religious, educational, social and recreational programs and activities. The building contained two swimming pools, showers, a full-sized basketball court

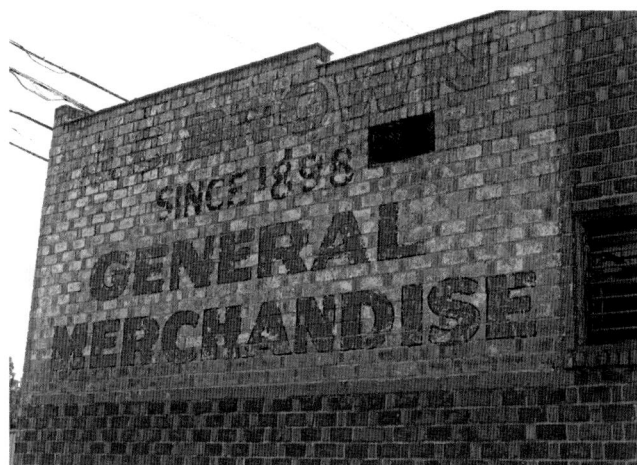

Right: J.C. Brown General Merchandise, West Huntsville. *Author photo*.

Below: The West Huntsville YMCA. *Courtesy of Larry Lyons.*

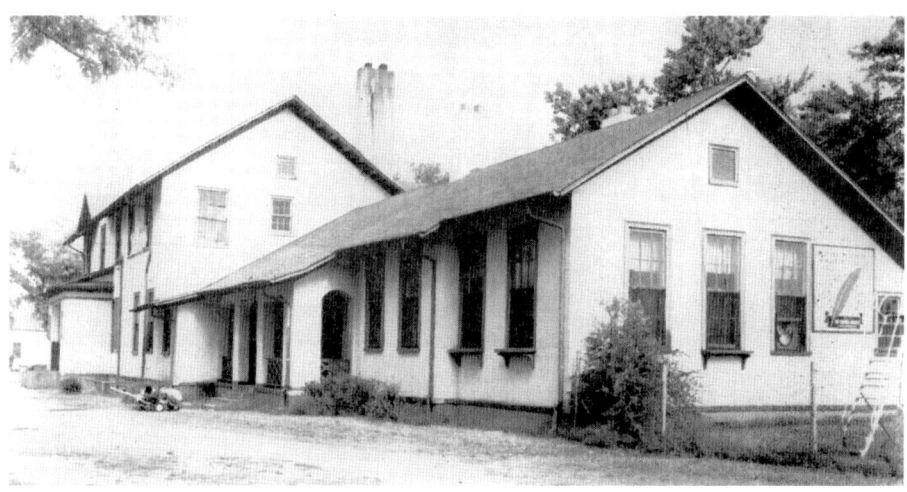

and a bowling alley. The McCormick Y was the hub of the community and at one point had over two thousand dues-paying members.[108] In their 1925 yearbook, the association's executives declared just how much influence they had within the West Huntsville community:

> *The Chamber of Commerce of Huntsville recently compiled a textile survey of the Huntsville district. This survey included West Huntsville, and one of the outstanding features of the survey was the fact that the textile mills of West Huntsville had never suffered a general strike, which was evidence conclusive that the labor of the district was a happy and contented people. Such conditions do not just happen. There is a reason. The West Huntsville*

Y.M.C.A. has played a very important part in the progress, happiness, and success of the industrial enterprises of the district. Their work has been reflected in the homes and industries of this thriving community.[109]

While this was true at the time, it would all change with the International Textile Union's strike of 1934, when Lowe Mill workers joined other Huntsville millworkers in walking off their jobs.

THE SCHOOL

None of the mill executives in West Huntsville, including those at Lowe Mill, were willing to undertake sole responsibility for providing a school for mill children. However, through the efforts of a Presbyterian missionary, Jessie House, and a young teacher, William P. Fanning, the mills were persuaded to donate a four-room dwelling for the first school, located on the corner of Sixth Street and Eighth Avenue. Then, in 1916, Virginia McCormick, founder of the YMCA, contributed the bulk of funds to construct an eight-room wooden building, with the remainder of the money coming from mill executives and the community. This structure, located on Ninth Avenue, sufficed until the county built a larger, more substantial brick schoolhouse on the same site, named West Huntsville School.[110] This school only went to the ninth grade until 1934, after which an additional grade was added each year. The first high school graduating class was in 1938. A new school building was erected on Clinton Street in August 1944.[111] Much of the success of the high school can be attributed to its principal, J. Homer Crim, who came to the West Huntsville community in 1933. Fourteen high school classes graduated from the school under Crim's firm guidance, the last in 1951.

In 1950, the principals of Huntsville's four mill schools—J. Homer Crim of West Huntsville, Cecil Fain of Rison, Elmon Brown of Lincoln and E.F. Dubose of Joe Bradley—met with Madison County superintendent Edward Anderson to discuss consolidation of the four county-run mill schools into a single county school. They decided the new school should be located at a point of interest nearest the center of student population—the West Huntsville district. On July 1, 1951, the Madison County School Board passed a motion to rename the West Huntsville High School after former school superintendent S.R. Butler.[112]

Huntsville Textile Mills & Villages

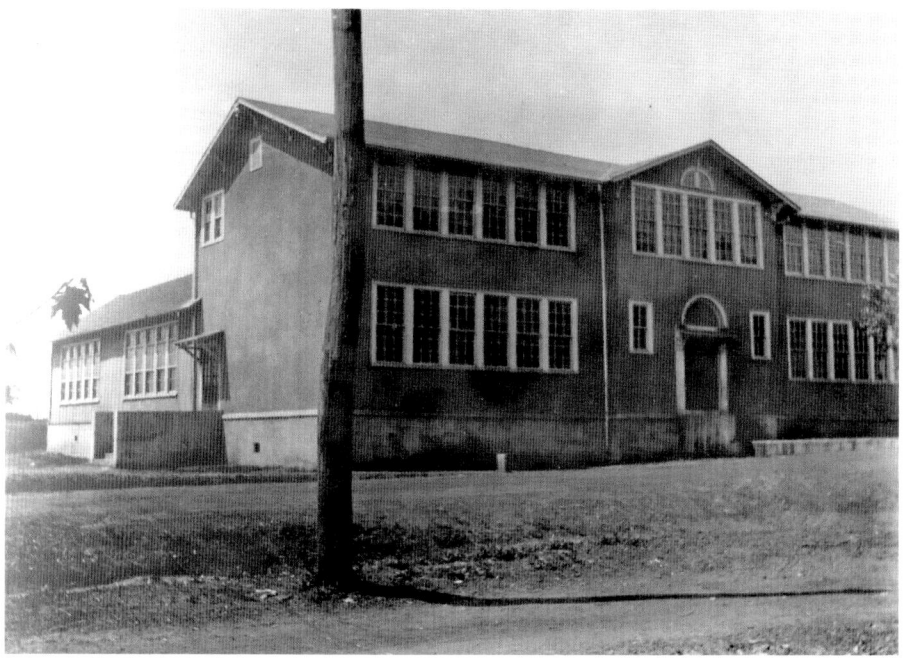

West Huntsville High School on Ninth Avenue closed in 1951 when mill students moved to the new S.R. Butler High School at the intersection of Clinton Avenue and Governors Drive. *Huntsville–Madison County Public Library Archives.*

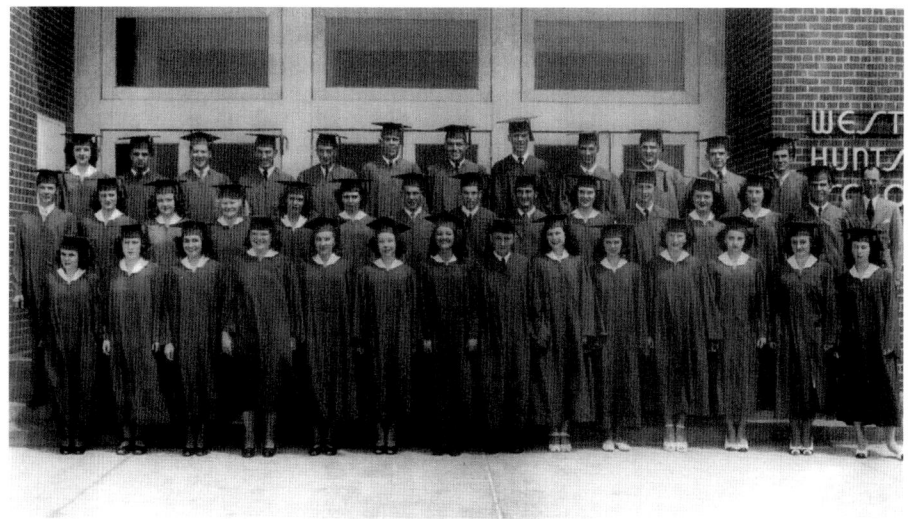

The West Huntsville graduating class of 1947 was the largest and most influential. The class students organized the first football team and produced the only school annual. *Courtesy of Betty Owens.*

S.R. Butler High School remained a county school until 1956, when it joined the city school system. The school was also the site of the University of Alabama's first extension center in the city in 1950, before the University of Alabama in Huntsville was formally founded in 1961.[113] In 1967, S.R. Butler moved from Clinton Avenue to a new building on Holmes Avenue, and the former building became Stone Middle School. Homer Crim would serve as Butler's principal until his retirement in 1971. Betty Owens was in the first graduating class at Butler High School in 1952. She remembers Homer Crim as a small man who ruled with an iron fist but had a heart of gold:

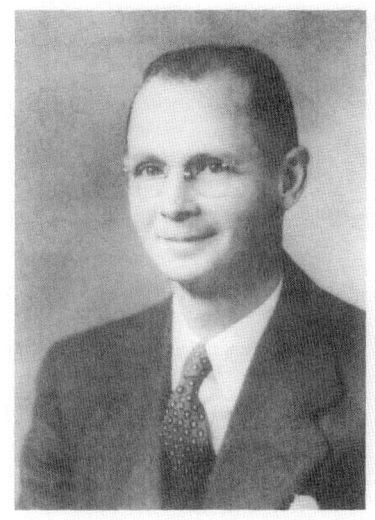

Homer Crim, principal of West Huntsville and S.R. Butler High School until his retirement in 1971. *Courtesy of Betty Owens.*

During the Depression a lot of people didn't have food to eat. Mr. Crim worked at Dunnavant's shoe store downtown and was principal and his wife taught school at West Huntsville during the day. He would go from house to house and ask for just a hand full of beans or whatever they might have. We had a building behind the school with a little kitchen and he would boil everything up and make soup to feed the children who didn't have anything to eat. He was a wonderful man.[114]

S.R. Butler High School grew in the 1970s to a student body of almost three thousand, integrating mill and factory children with the army children of Redstone Arsenal and the children of some of Huntsville's first engineers and technicians.[115] Former student and *Huntsville Times* writer Lee Roop likened it to "taking classes in an international airport." Butler also integrated black and white children. Racial tension became a part of Butler's identity, but it was also a school known for its academics, music and sports teams. "They say America is re-segregating," said Roop, "and I hope that's not true. I hope Huntsville always has places like Butler where we meet and talk to people who aren't like us."[116] The school, which merged mill kids, army kids and all races, closed its doors in 2015.

7
THE EAST HUNTSVILLE ADDITION

Though not actually a mill village, one would be remiss to not mention the area known as the East Huntsville Addition and its importance to the lives of mill families. It was Huntsville's first suburb and a reflection of the city's industrial growth at the end of the nineteenth century and into the twentieth century.

In 1888, the NAIC made a real estate venture and created the East Huntsville Addition in an effort to attract industry and jobs during the Reconstruction period. The company procured two thousand acres of land from individual owners. The plat for the property was approved on February 22 of that year. The following May, lot sales were kicked off with a day-long celebration and barbeque. Local residents C.H. and W. Halsey purchased the first lots for $300. Despite this grand gala, sales did not immediately take off:

> *Even though the Improvement Company stockholders had worked hard promoting Huntsville, the economy did not begin to soar until the early 1890s when many new industries became fully established, among them the textile mills, which were to be the cornerstone of prosperity for many decades.*[117]

The NLA, consisting of the troika of Wells, Wellman and Pratt, along with James A. Ward, bailed out the NAIC and purchased almost the entire East Huntsville Addition, except for a western portion that the NAIC had deeded to Dallas Mill, which was then under construction. In 1892, a 285-acre tract was replatted and subdivided into residential lots. The new plat called for a large boulevard, aptly named Pratt Avenue. In fact, the street names running east and west read as a who's who of prominent Huntsville citizens and association members, while the side streets at the time were merely numbered.

To entice additional investors and homeowners, the NLA printed thousands of promotional brochures, which it sent all around the country. When the promotion ended, Pratt and his comrades helped to establish the first Huntsville Chamber of Commerce in 1894 with Pratt was president.[118] Lots sold rapidly under the NLA's management, and construction began at once on new houses. Both Wells and Wellman owned homes in the new district. In addition, in 1892, Wellman purchased $63,000 worth of East Huntsville lots. The textile mills secured the success of East Huntsville, especially with the addition of the streetcar line to connect the mills and surrounding housing with downtown.

In 1999, the city created Five Points Historic District between Ward Avenue on the north and Wells Avenue on the south, Russell Street on the west and Grayson Street on the east. About thirty-five homes now have markers identifying the original owners. Although the historic district does not cover all of the area known as Five Points or the entire East Huntsville Addition, it does encompass prime examples of the district's architectural diversity, including English cottage, Colonial Revival, duplexes, Victorian and, more recently, ranch homes. The mélange of residential structures is representative of how Huntsville grew and changed from the cotton mill era to the space age of the 1950s and '60s. The area also has a very eclectic business district, including arts, antiques and a coffee house whose former owner, Cheryl Sendowski, was the daughter of a millworker. "The appeal is, it takes you back in time," said Sendowski. "It's that feel, that Norman Rockwell feel."[119] That "feel" contributes to making the Five Points Historic District a true slice of southern Americana.

8
DANGERS, UNREST AND UPHEAVAL

RISKY BUSINESS

While the amenities offered within the mill villages made life more comfortable for workers than it had been in the mountains and on farms, it was by no means paradise. Neither were working conditions within the southern mills completely hellish. To paint such an extreme picture would be inaccurate and unjust. The palette of village and factory life was not black, white or even gray but one that spans the spectrum of colors. That being said, there were dangers associated with working in the mill, and discriminatory practices were tolerated for the sake of production and profit.

Mill work was dangerous. In each mill, a master switch controlled all machinery. Shutting down one machine shut them all down; thus, management was reluctant to stop the machines and bring production to a halt, even in the case of an accident. It only took a split second of carelessness for a worker to lose a finger, hand or arm in machines designed to shred and tear at cotton fibers. Often the loud machinery made it difficult to hear, which could also result in injury on the job and permanent hearing loss. Workers sometimes sought compensation for their injuries, but almost always the mills won such cases.

Sometimes freak accidents would occur, like this one recalled by Dallas Mill village resident Sara Ann "Sally" Certain Hymer:

> *In October 1928, my dad, Lionel Certain, who had gone to work at Dallas Mill, was walking down an aisle in the mill when a bobbin fell from a top*

rack and hit him on his forehead at his hairline; it did not break the skin. He developed a headache and told his supervisor (who was his brother, Clarence) what had happened and Clarence told him to go home.

When he went to the doctor the next day because of the severe headache, the doctor said he had a blood clot on his brain. From that time on he suffered severe headaches and had two operations to remove pieces of his skull to relieve pressure on his brain. He died on March 3, 1929, at Huntsville Hospital after the second operation; he was 19 years old; I was 5½ months old.[120]

Another hazard was the thick cotton dust that constantly filled the air. Not only did this dust settle on workers' hair and clothing but also in their lungs. Workers called this "eatin' cotton"; doctors call it byssinosis, or "brown lung" disease. Byssinosis is an old disease. Physicians in Europe described it as a respiratory disease in textile workers as early as the eighteenth century. In the United States, especially in the South, industry executives for most of the twentieth century said it was a "phantom disease" and attributed it to liquor and laziness. As late as 1981, Thomas G. Auchter, then administrator of the safety and health agency, which is part of the National Labor Department, ordered that all copies of a booklet issued by the health agency during the Carter administration be destroyed. While not questioning its content, which described the dangers of inhaling cotton dust, he proclaimed the cover of the booklet, which portrayed a worker obviously ill with byssinosis, to be "biased."[121] Clearly, big business and unscrupulous government officials have always been able to deny and skew facts in their favor.

CHILD LABOR

The flying lint was even more dangerous to children's delicate lungs. Children often accompanied their parents to work and, when they were "old enough," worked alongside them. Child labor was cheap and some say a "necessary evil" of the quickly expanding industrial age. Employers were not altogether at fault in hiring children. Parents often insisted that employment should be given to the entire family, and a manager in need of operatives would have been thought foolish to reject the whole family merely because some of the children were young.

Child labor was a "special curse" of the cotton mill industry—first in England, then in New England and later in the South. "It may be broadly stated that there would be no child labor problem to speak of in the South today except for the cotton mill, and this industry is centered in the piedmont section of four cotton growing states—North and South Carolina, Georgia and Alabama,"[122] said A.J. McKelway, a North Carolina clergyman, social reformer and journalist who wrote a series of editorials in the early twentieth century opposing the labor of young children in the mills.

In 1887, Alabama passed a law forbidding any child under fourteen to work more than eight hours a day in a mill, but then an influx of northern capital came into the state to be invested in the cotton mill industry. Through the influence of the northern mill owners, the law was repealed, and through the same influence—supported by southern manufacturers—the child labor law proposed in Alabama was repealed in 1895.[123] In the wake of this repeal there was much agitation and an increased effort to secure the law's passage. "Southern patriots everywhere are proclaiming that the child should be put above the dividend," wrote McKelway. "The place for the child is not the mill, but the school."[124]

In 1903, Alabama adopted its first child labor law, but little was done to implement or enforce it. A Huntsville newspaper reported that the

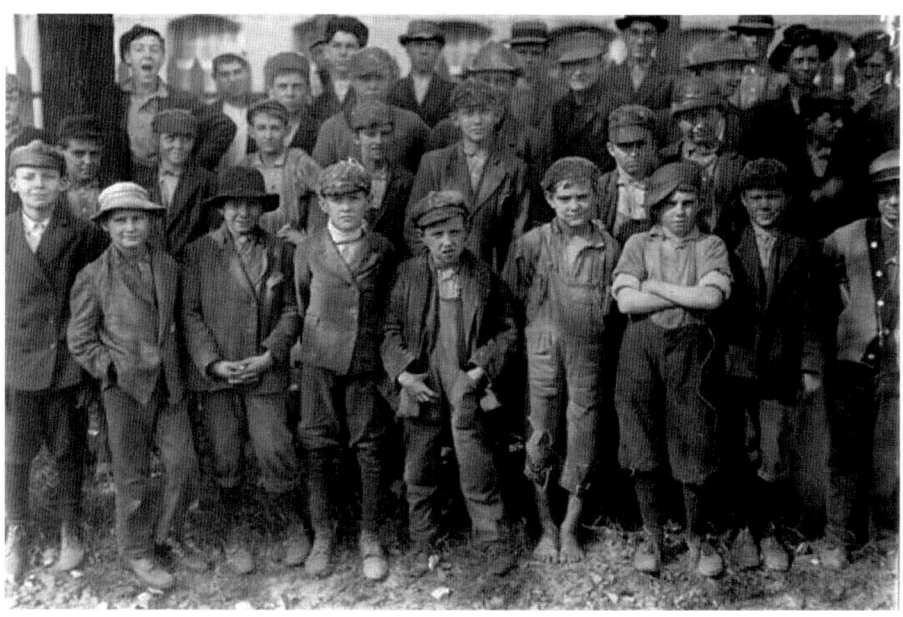

Child workers at Dallas Mill, 1910. Photo by Lewis Hine. *Library of Congress.*

law would compel three hundred children to quit working in local mills. This was emphatically denied by William R. Rison, general manager of the Dallas Mill company, who stated that it was against the rules of the company to employ any child under twelve years of age, and if there were any younger operatives, their ages had been misrepresented by their parents.[125] The law did not allow a child under twelve to work unless he or she was a widow's child.

The Reverend Neal L. Anderson, a member of the Alabama Child Labor Committee out of Montgomery, visited two factories in Alabama and reported his findings:

> *Out of at least a score of children, evidently under the legal age limit, twelve, in a mill employing some three hundred operatives, only one child was found who was not over twelve.... This child confided to a little girl who was in my party the information that she was ten years old, and she looked younger still. I asked a little boy, who could not have been over nine or ten years old, how old he was. He replied with a wink and a roguish laugh that he was "most fourteen," and then ran off to tell the other children that the stranger wanted to know his age.*[126]

The National Child Labor Committee had been trying to enforce child labor laws since it was founded in 1904. In 1908, it enlisted the help of photographer Lewis Hine to get the message out. For the next decade, Hine traveled to half of the continental United States taking photos. He usually visited incognito, posing as a Bible salesman or an industrial photographer making a record of factory machinery.[127] Hine came to Huntsville three times, first in November 1910 and, later, during two separate visits, in November and December 1913. He was denied access in both Dallas and Merrimack Mills and had to capture photos of the children after work or near their homes. He took about thirty pictures in Huntsville over his three visits. Smiling Charlie Foster's photo was snapped by Hine in 1913, as he was leaving the Merrimack factory. Charlie became a poster child for issues related to child labor laws.[128]

One of the best methods of forcing compliance with the child labor law was to have a compulsory education law. While Alabama law at the time of Hine's visit did require children to attend school for eight weeks out of the year, absenteeism was high, and there was little provision for inspection or enforcement. The state had created a department for the inspection of jails, factories and almshouses, but it had only one inspector and one assistant and,

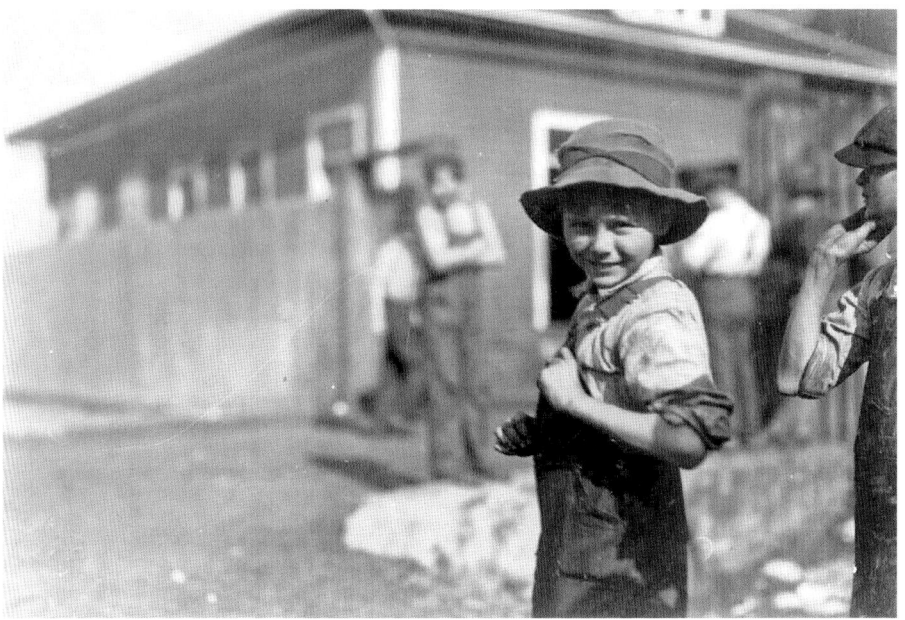

"Charlie Foster has a steady job in the Merrimack Mills. School Record says he is now ten years old. His father told me that he could not read, and still he is putting him into the mill." Photo by Lewis Hine. *Library of Congress.*

as McKelway reported, "On account of the illness of the present inspector and of his predecessor, very little has been done."[129]

Reports indicated that a number of mills did drop children from the payrolls, but they were often still employed as "helpers," one of the most frequent ploys to evade the law.[130] Neither did it help that many uneducated parents saw education as a waste of time when a child could be working and helping to support the family. "My mother grew bitter toward many of 'those old farmers' who sat at the mill gate and took the pay from their children, spat tobacco juice and did no work. My grandfather and father were not a part of that scenario," said former Merrimack villager Bill Brown. Even those parents in favor of education expected their children to earn their keep. "Mother said on Fridays after they'd get paid she'd hand an envelope with money in it over to her daddy at the supper table," said Doris Brown. Her mother started working at the mill at age nine or ten. "She was so tiny," said Doris, "when they came through to check for the labor law they'd hide her under the steps."[131]

Linthead Legacy

The illiteracy of the manufacturing states of the South is largely due to the competition for the life of the child between the school and the mill, with the manufacturer too often, the parent nearly always, and sometimes the child, on the side of the mill.[132]

In 1914, former state superintendent Henry J. Willingham gave this passionate plea to the people of Alabama regarding the education of their children:

School attendance is required throughout the civilized world to-day, except in Russia, Spain, and Turkey, and six of the Southern States. How much longer shall we in Alabama be willing to say "Here we rest"? It is gratifying to observe that public sentiment seems to be crystalizing in the demand for a law upon our statute books on the subject of compulsory attendance. God speed the day when it may come.[133]

By 1915, Alabama law prohibited children under sixteen from working with dangerous machinery. Fines for breaking the law ranged from fifty to one hundred dollars, but enforcement was difficult. Women's reform efforts reduced the rate of child labor in textile mills from nearly half of all workers in 1900 to a quarter of all workers in 1920.[134]

By the 1930s, child labor had already begun to decrease, as the Great Depression forced adults to compete for the low-paying jobs usually held by children. The active struggles against child labor also tended to strengthen unionism and workers' rights in general, as union leaders made their way from New England to the southern states. In Alabama, the apprehension of the mill owners concerning a child labor law and compulsory school attendance was tempered by the threat of unionism. It was pointed out that nothing would strengthen the cause of labor organizations in the South as much as their alliance with the cause of innocent, helpless children against capitalist oppression. By 1938, the minimum age of employment and hours of work for children became regulated by federal law.[135]

To its credit, the South did act much more quickly to solve the problem of child labor than did England, which took one hundred years to pass a law raising the minimum age limit to twelve years. And it speaks well of Huntsville that men like Big Joe Bradley supported education and helped to build first-class schools and bring in dedicated teachers for the mill village children. But worker unrest in Huntsville and all over the South—exacerbated by the Depression—continued to rise and was about to cause major upheavals throughout the region.

The Strike of 1934

Huntsville mills found their warehouses glutted as the wartime boom for cotton goods ended and foreign competition began to cut into the U.S. markets. Workers found themselves without jobs, and those remaining were expected to work multiple shifts with limited breaks and operate more machines for the same low wage. Workers referred to this unfair labor practice as "stretch-out," as it literally stretched them to their limits.[136]

By 1933, President Franklin Delano Roosevelt's New Deal environment yielded the National Industrial Recovery Act (NIRA), which protected the rights of workers to organize, and the National Rights Administration (NRA). The NRA established provisions that set a two-year suspension of the federal antitrust laws, enabling business leaders to draw up their own industry-wide "codes of fair competition," the right to join a union unhindered and the requirement that employers agree to provisions setting maximum working hours and minimum wage rate.[137] Business leaders felt the labor provisions "sufficiently vague as to minimize the prospect of their having to bargain seriously with employees' unions."[138]

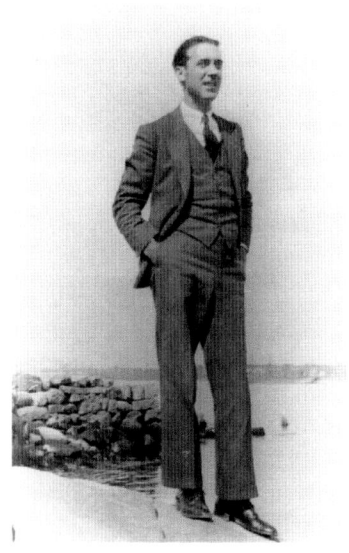

John Dean, United Textile Workers of America organizer from New York assigned to Alabama, 1935. *Huntsville–Madison County Public Library Archives.*

H.D. Lisk, a southern organizer, believed the NRA was something "God had sent to them." "I sure am proud," said one Alabama worker, "of our president as it is the first time the laboring class of people has anyone to help them."[139] Within a year, membership in the nation's textile mill unions went from 40,000 to 270,000.[140] Millworkers knew that all those mill amenities came with a price—strict adherence to mill rules, sober and moral living and, they were soon to discover, no contact with labor unions. The northern owners and supervisors in the Huntsville factories had had their fill of unions while in the North and had no desire to take up the battle again. Mill church ministers, as employees of the company, often aided owners in their cause by focusing their sermons on the virtue of hard work, integrity and sobriety. A worker suspected of consorting with union activists was in jeopardy of being fired and he and his family evicted. The mill housing perk provided mill owners with a great leverage tool against union organization.[141]

Because of the growth of membership in Alabama, United Textile Workers of America (UTW) president T.F. McMahon sent John Dean, a veteran UTW organizer, from the Northeast to assist in union enrollment and harness the mounting frustration brought on by mill owners' and managers' refusal to comply with NRA standards. Dean, along with UTW state official Albert W. Cox and experienced organizers Mollie Dowd and Alice Berry, began a campaign to address workers' grievances. State UTW headquarters was located in the Russel Erskine Hotel on Clinton Avenue. Concerns included eliminating the stretch-out, establishing a twelve-dollar-per-week minimum wage and a thirty-hour workweek, reinstating any worker fired because of union organizing and recognition of the workers' right to organize.

There is some discrepancy as to where the strike actually began. One source names Huntsville as the location of the first walkouts, when four thousand workers left their jobs:

> *As the southern state with the strongest tradition of militant unionism, it is not surprising that the largest strike to that time in American history should have begun in Huntsville. But it is a piece of Alabama history curiously forgotten by most of its population.*[142]

Another source states that workers at Dwight Mills in Gadsden, Alabama, walked out two days before Huntsville workers.[143]

In either case, angry north Alabama operatives were unwilling to wait for the national UTW union to call a strike and staged a walkout in mid-July 1934. Of the forty-two Alabama UTW locals, forty voted to strike. The South's fervor spread, and the pressure was felt to call for a nationwide strike, which union vice president Francis Gorman reluctantly called for on September 1, 1934. Eventually, the strike would include 400,000 textile workers from Alabama to Maine.[144]

Southern textile workers put tremendous faith in the NRA to end the stretch-out and raise wages, but the agency rarely, if ever, lived up to its promises and provided no effective means to enforce its provisions. Dallas Mill workers met at a local church, where Monroe Adcock, president of the Dallas local union, urged that no destruction of mill property take place during the strike and that millworkers refrain from imbibing in "intoxicating liquors."[145] Flying squadrons of armed men piled into old clunkers, driving

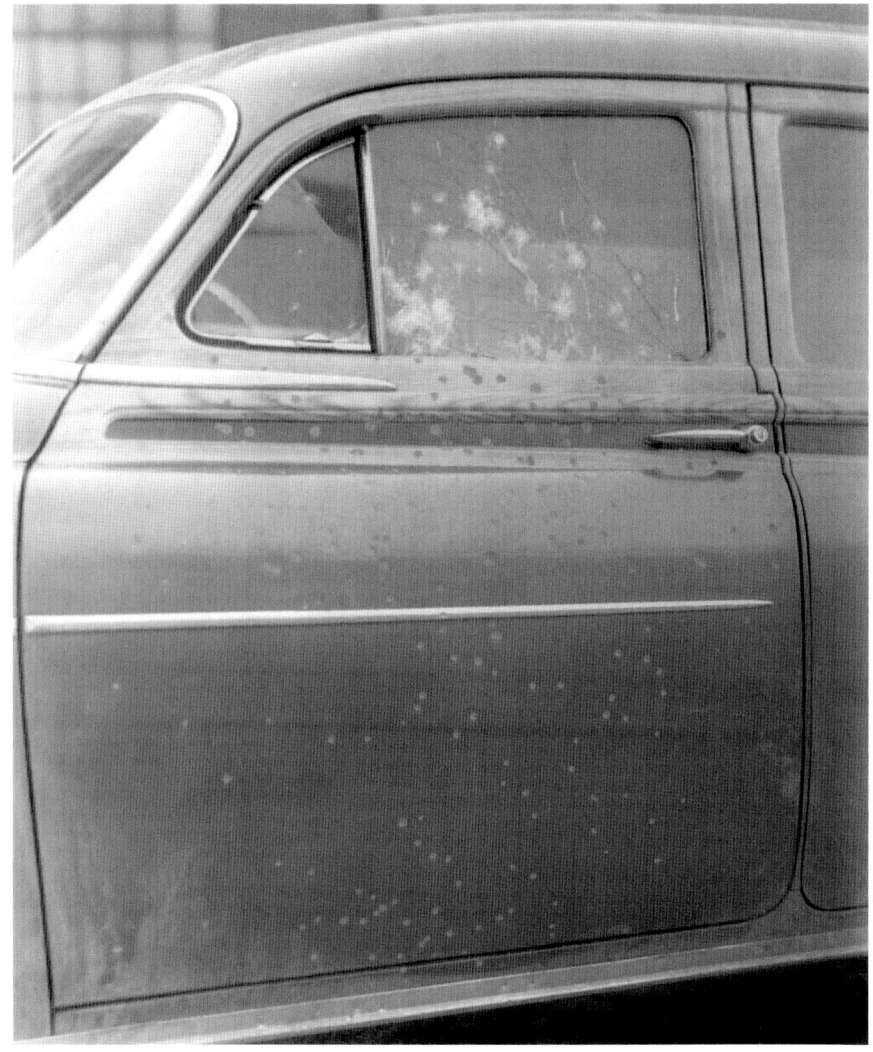

A car riddled by gunfire during the 1951 HIC strike. *Huntsville–Madison County Public Library Archives.*

from mill to mill to compel other workers to join their ranks. Compared to some of the other southern states like the Carolinas and Georgia, Alabama kept the cause passionate yet restrained. One incidence of violence occurred in Decatur, where Huntsville organizers were shot at.[146]

The most sensational event in Huntsville was the kidnapping of John Dean. Dean had been very public in his rallying of Huntsville's UTW members.

He'd appeared in Huntsville's Labor Day parade, where six thousand men, women and children were in attendance, waving from an open car flanked by officials of the Huntsville Trades and Labor Council. He spoke outdoors to employees of Lincoln and Merrimack Mills and also to several hundred Dallas Mill workers at a local Methodist church. Dean roused the workers to great enthusiasm when he told them the United Textile Workers of America was behind the Alabama movement and would see that the strike was "carried out to success."[147]

Not everyone was taken with Dean's fervency. Early Sunday morning, August 6, 1934, the union organizer was kidnapped from his room on the sixth floor of the Russel Erskine Hotel by four armed men. His abductors drove him to Fayetteville, Tennessee, roughing him up along the way. Strangely, they checked him in to a hotel and left him. In less than an hour, a cavalcade of armed men came to rescue their leader. Strikers wanted to take matters into their own hands and take Dean's abductors captive. Armed road blocks were set up, and the men demanded that the city take action. Though no one was yet named, the district attorney assured the men that a warrant had been issued. Dean refused to remain sequestered.

> *Monday morning found a large crowd assembled downtown awaiting the day's event. In an act of bravado, Dean drove in from Merrimac [sic] and casually breakfasted at the Central Cafe downtown while armed bodyguards patrolled the sidewalks out front.*[148]

On August 13, a kidnapping charge was issued against James Conner, a millworker. That charge was later lessened to "whitecapping," defined as "an act to prevent and punish the formation or continuance of conspiracies and combinations for certain unlawful purposes." The trial was set for November 28, 1935, but was continued until February 19, when the entire matter was, for some unknown reason, dropped.[149]

In September 1934, Roosevelt reorganized elements of the NIRA to create an independent review board not held captive by mill owners. This new board did nothing to alleviate the stretch-out and wage system, and workers' lives went on much as they had. By September 22, Gorman had called off the strike. What the national leadership was deeming a victory was really a defeat. Bitter and disillusioned workers returned to the job, some to be let go or not allowed back at all.

"Given the lack of any national or even statewide overview, it is hard to even call what happened in Alabama a strike at all, but rather

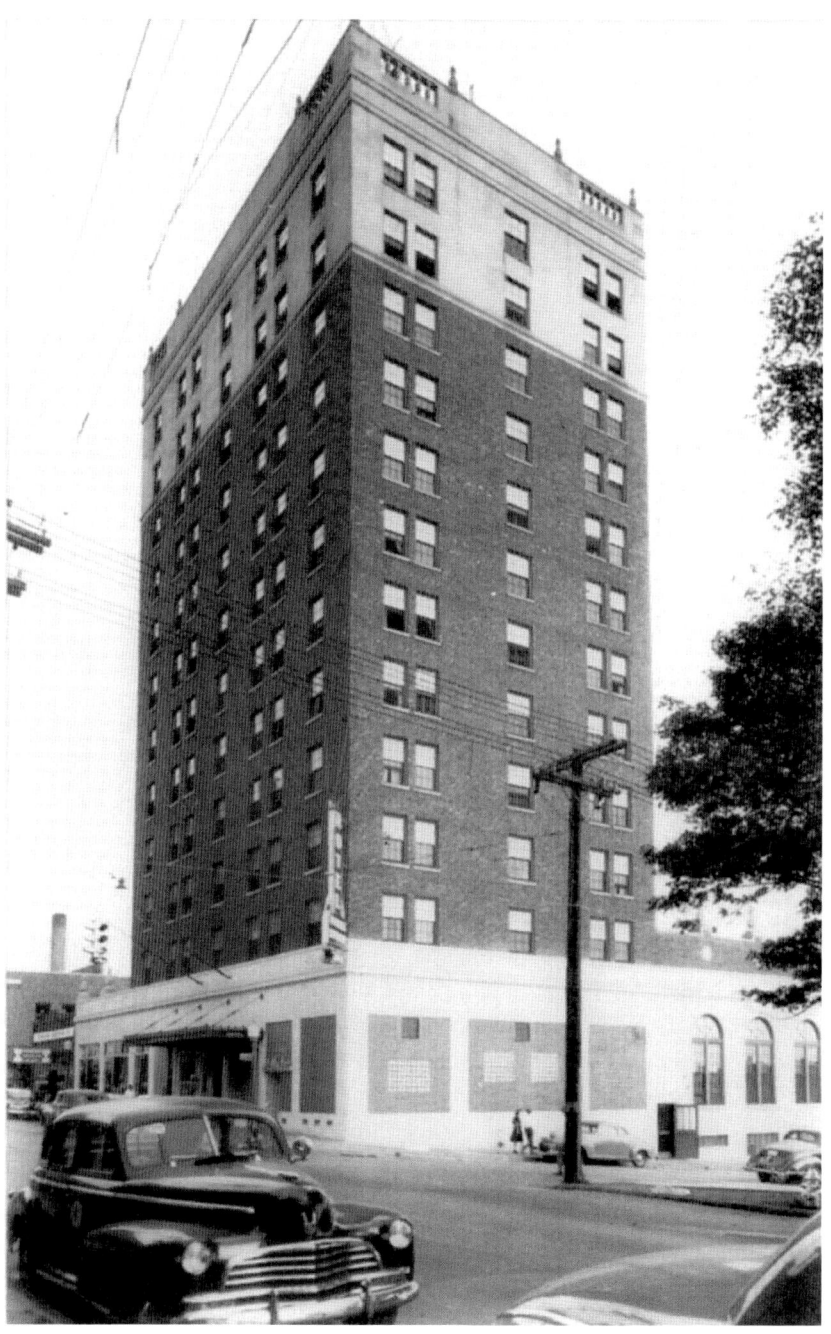

The Russel Erskine Hotel *Courtesy of Larry Lyons.*

a succession of individual walkouts, instigated by local leadership," said historian John A. Salmond.[150] The strike of 1934 dealt Huntsville a severe blow; the cotton mill industry struggled in its aftermath and throughout the remainder of the Great Depression. By 1955, only Merrimack Mill remained in operation.[151] The era of the great textile industry was all but extinct in Huntsville, Alabama.

9
PRESERVATION, RESTORATION AND REVITALIZATION

Preservation Versus Progress

It's an age-old dilemma—preservation versus progress—how to maintain the historical integrity of a city while allowing for the cultural, structural and technological evolution that is destined by time. In 1950, five years before the last textile mill's doors closed for good, German-born rocket engineer Wernher von Braun came to Huntsville to head the U.S. space program at Redstone Arsenal. Huntsville had progressed in short order from a cotton mill city to "The Rocket City." Some mill villagers, like Merrimack Mill employee Bill Brown, even went on to work on the arsenal. "The people who worked in this mill provided the base for the people who went to work for Wernher von Braun," said Merrimack Hall founder Debra Jenkins. "Their parents had worked in the mills, but they had gotten educations in math and science and were able to get jobs that weren't just manual labor at the textile mill."[152]

By 1956, the city limits of Huntsville included all of the mill villages. Some mill villagers kept their homes, having purchased them from the mill owners. This allowed them a sense of freedom and individuality like never before. People painted their homes and made other minor changes that altered them from the cookie-cutter shotgun houses and duplexes built by the mill owners. While former millworkers and their families were officially citizens of the city of Huntsville, there remained a palpable separateness, and there was little community support to preserve the villages. Lauren Burlison Martinson had this to say in her 1998 master's thesis, "Revitalization and Preservation in Alabama's Textile Mill Villages":

Linthead Legacy

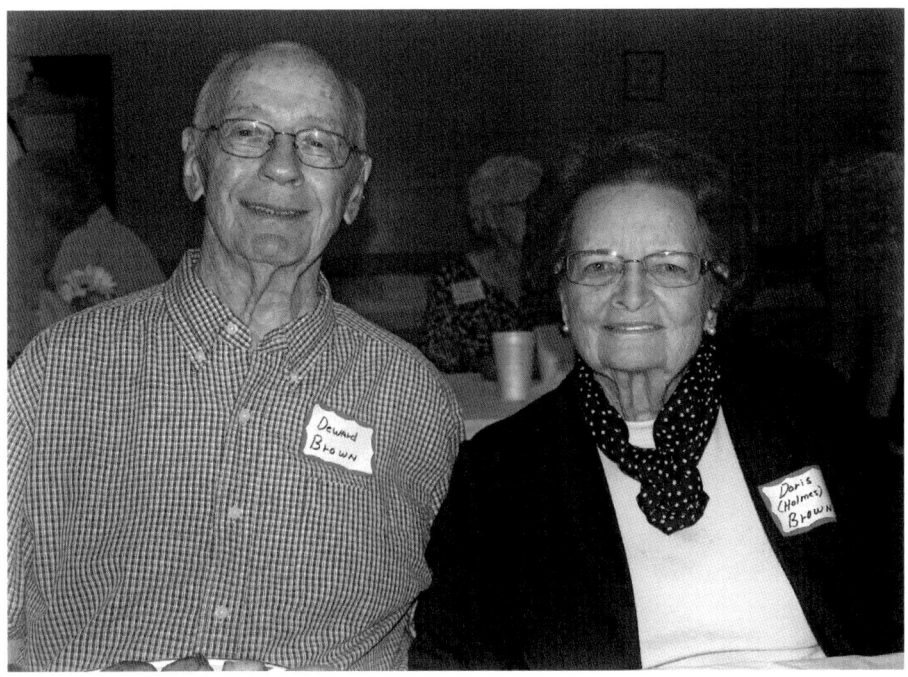

Deward "Bill" Brown and his wife, Doris (Holmes) Brown, at the Merrimack Mill/Joe Bradley reunion, June 2016. *Author photo.*

Mill villages have historical and current perceptions that tend to work against them. From a historical perspective, mill villages are often reminders of exploitation and control where second class citizens eaked [sic] *out a living performing menial jobs in a dehumanizing, industrial setting. Today the neighborhoods are looked upon as areas not worth saving because all that appears is a legacy of disinvestment, deterioration and social problems that seem on the surface to be too large to solve. What fuels these perceptions is a lack of understanding of the true historical significance of the mill village and the absence of a unified group of people working together to change the way mill village communities are perceived.*[153]

In the twenty-first century, perhaps enough time has passed that people's perceptions and priorities have begun to change. Many local and state government boards and commissions, community organizations and businesses are now working together to preserve these historic districts. Every mill factory, village and some of the schools and structures within those villages is now in the National Register of Historic Places, the nation's official list of historic buildings and other cultural resources considered to be worthy of preservation. Lincoln Mill and Merrimack Mill were added to the register in May 2010 and Dallas Mill and Lowe Mill in June 2011. The Alabama Historical Commission is the preservation agency of the State of Alabama that nominates properties to the National Register. That nomination must go through a state review board and the state historic preservation officer before it is approved by the National Park Service. "The Historic Huntsville Foundation invested over $12,000 to have mill villages placed on the National Register of Historic Places—in addition to the Certified Local Government funds invested by the Alabama Historical Commission," said Donna Castellano, executive director of the foundation.[154] Individuals, businesses, churches and various organizations have sponsored historic markers, which are placed on the sites by the Huntsville–Madison County Historical Society.

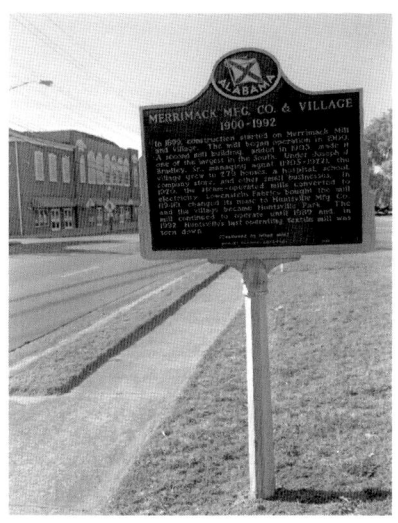

Individuals, businesses, churches and various organizations have sponsored historic markers placed on the sites by the Huntsville–Madison County Historical Society. *Author photo.*

"While National Register designation helps to provide visibility to historic districts, it is just that, a designation," says Jessica White, a historic preservation consultant with the City of Huntsville. "National Register designation does not protect buildings from change or demolition….If we start chipping away at that historic integrity over time we'll lose the value that historic places can provide us with."

In order to be protected from change or demolition, the homes must be recognized by the City of Huntsville as historic districts. While all of the mill districts now have National Register status, they are not recognized by the city as historical districts. According to White, the city requires that 60 percent of the property owners in the historic district petition city council to become a locally designated historic district. Local designation means that properties in those districts would come under the purview of the Historic Huntsville Preservation Commission (HHPC) and be subject to a design review process. White says that it is likely that many of the landlords and property owners in these districts do not wish to be held to these standards:

> *One means of protecting our districts lies in local designation. Through local designation districts can be better protected from the ravages of extreme alteration and demolition. Local designation involves the oversight of a Historic Preservation Commission or an Architectural Review Board. These overseeing bodies typically follow local guidelines, which are based off of the Secretary of the Interior's Standards for Rehabilitation and are enforceable by local ordinance. While these organizations aren't the be all end all answer for preserving buildings, they do go a long way in preventing a total loss of historic material.*

However, White adds that maintaining the integrity of the mill districts is more than working to preserve and restore its buildings:

> *Historic places intrinsically attract people because they are a tangible record of our heritage, providing a much needed link to our history. Historic places are our roots, they provide us with a sense of place and solidarity....Another thing to keep in mind when preserving districts, like the milling districts in Huntsville, is that we shouldn't just be in the business of saving buildings, we should also focus on saving the communities that helped to build them. Oftentimes it's people who make places unique and intriguing. We should be weary [sic] of the cultural shift that can happen through gentrification. Maintaining authenticity and community spirit is key to maintain historic integrity in our neighborhoods.*[155]

Dallas Mill District

Court Heller, a local architect and treasurer of the Northeast Huntsville Civic Association (NEHCA), hopes Dallas Mill village's National Register listing will bring in homebuyers and help offset some of the loss associated with construction of the Interstate 565–Oakwood Avenue interchange. The old Rison School was razed to make room for the elevated section of highway. Construction, fire, neglect and absentee landlords have all contributed to the deterioration of the historic district. Heller remains optimistic that the area can be resuscitated:

> *There's a whole movement now of these tiny houses. Some of these homes would be great for someone out of college with a good start-up job who*

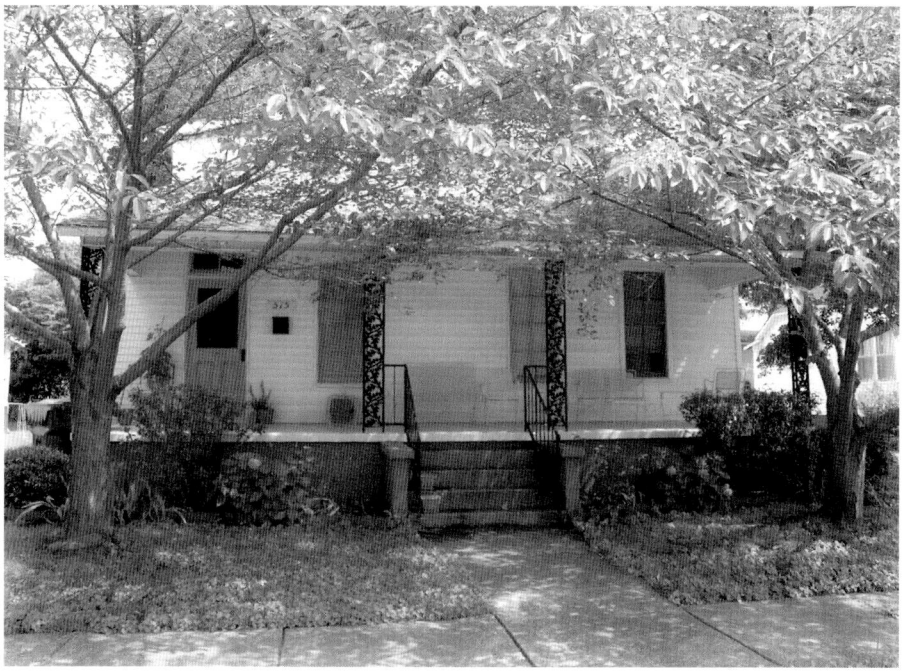

Home of Court and Amanda Heller on Humes Avenue. The former mill village duplex has been renovated into a single-family dwelling. *Author photo.*

> *wants something to live in that they could work on at night and make their own.... What I like about Dallas and some of the other mill villages is they have what I call "random uniformity." They were built to look the same a hundred years ago, but they've been modified differently by individuals over time....*
>
> *The problem with trying to save an area like this is you're trying to save a historical fabric but all of the little buttons on the blouse are completely different. You just can't get all those buttons to be the same....Individually there is only so much you can do. You can't turn these into McMansions like what's happening in some other parts of town.*

Heller, his wife, Amanda, and their brood of cats have lived in two former mill houses, the most recent a century-old refurbished duplex located on Humes Avenue. He said he purchased their latest home primarily for the massive pecan tree in the backyard. "The lady I bought it from, she and her sister planted that tree when they were children. She still asks about her pecan tree," said Heller. He says neighbors know and look after one another.

"It's still kind of like it was in the old mill days." Heller said bringing in more arts and entertainment and family events will help to revive the area. His wish is for the neighborhoods to again be safe and vibrant areas in which to raise families.[156]

In 2007, the Huntsville City Council approved spending $200,000 to refurbish Dallas Mills' aging water tower in northeast Huntsville. The council chose to refurbish the tower as a historic landmark at the request of those in the community, including Heller. Huntsville Utilities managed the tank rehabilitation project, which included a new roof, reinforced rivets and lots of paint. Drivers going east on I-565 will read "Dallas Mill" on the tower, while those driving west will read "Lincoln Mill." "It was a nod to both mills, since Lincoln no longer had its tower," said Heller. While the structure is now technically functional, there are no plans to use it. "I still want to see that thing light up like Lowe Mill's tower," said Heller wistfully.

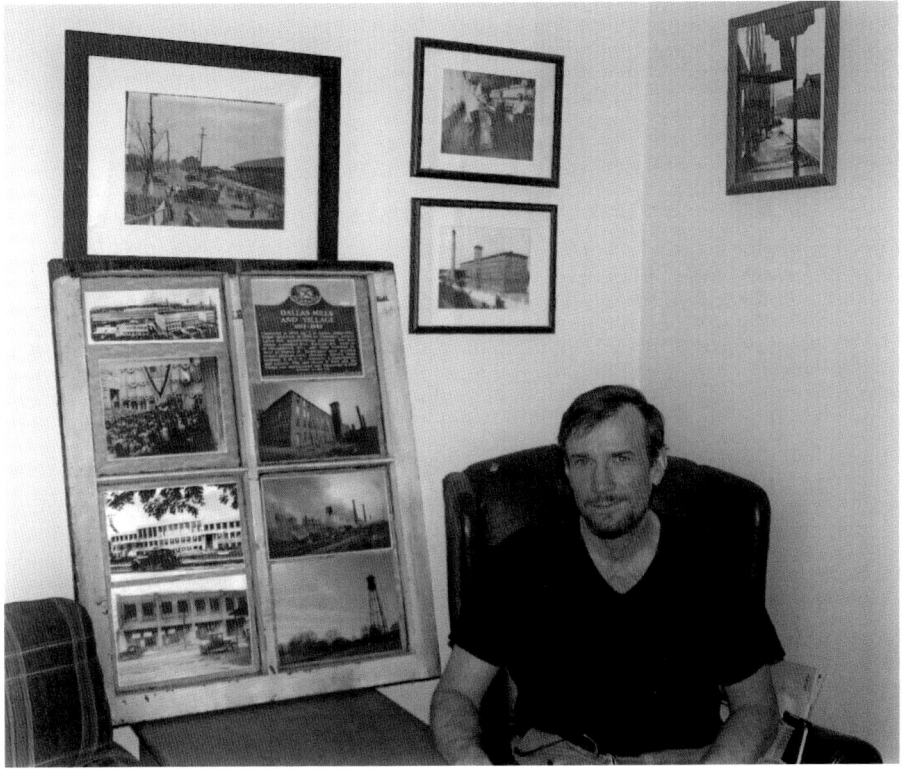

Court Heller's home on Humes Avenue is full of Dallas Mill photos. *Author photo*.

Goldsmith-Schiffman Field, located about one mile north of downtown on Beirne Avenue, was donated to the city by the Goldsmith-Schiffman families in 2014. Hundreds of high school football games were held on the site in decades past. The field will continue to be used for recreation league youth football, soccer and lacrosse. "Nothing could make the ancestors happier," said Margaret Goldman, great-granddaughter of Dallas Mill developer Oscar Goldman. "I think they're all clapping, wherever they are. I'm delighted that the people will continue to have wonderful memories of this park."[157]

In 1994, the City of Huntsville renovated and reopened Optimist Park, where Huntsville youth continue to play on its fields and in the recreation center.[158] Recently, the NEHCA, with the help of volunteers, strung one thousand peace cranes around the center, gifting some to the Huntsville Police Department. Each origami crane represents a prayer for good relationships with police and one another as neighbors, according to NEHCA president Frances Akridge. Recently, the NEHCA was the recipient of one of twelve micro-grants awarded by Downtown Huntsville Incorporated and PNC Bank. According to Akridge, the organization has plans to use the money for various events to bring the neighborhood together. Though Lincoln and Dallas Mill residents were once rivals, the districts are now part of the greater northeast Huntsville community. "We need to try to remember to break down those railroad tracks in our brain," said Akridge. "So many of our problems would be solved if we got to know our neighbors."[159]

Lincoln Mill District

When ophthalmologist Jim Byrne purchased the remaining Lincoln Mill buildings in 2007, he admitted he might have bitten off more than he could chew. "I could either be a huge fool, or a dreamer, or a successful investor. I guess time will tell," he said. Byrne and his partner, Wayne Bonner, took two to three years to clean out the building—essentially gutting it—polishing the seven-inch concrete floors, painting and adding air conditioning and more. The exterior of the building has been completely preserved, and though all of the window grids are original, every pane had to be replaced with insulated glass. "The windows were a pain in the butt," said Byrne. "There are thousands and thousands of windows and we've replaced every one at least once." Just like the children of yesteryear, kids today still like to get into mischief. "They stand on the railroad tracks and throw stones at the windows. It's a real thorn in my side," Byrne said. The steel and concrete framework is what saved the remaining structure from the fire that destroyed much of the factory in 1980. "Structural engineers went through the building. One told me he'd never been in a building this old that didn't have any settling, any cracks, any shifts. He said there wasn't one thing in this building that wasn't structurally sound," said Byrne. The complex located on Meridian Street, home to Wernher von Braun's first Huntsville office in the late 1950s, is now called Lincoln Mill Office Campus. An eclectic mix of tenants rent space in the building, from high-tech and defense offices to a men's clothier and a hair salon. The venue also hosts a massive costume party called Yuri's Night. The internationally recognized event held each April celebrates the first spaceflight achieved by soviet pilot Yuri Gagarin in 1961 and the voyage of the American space shuttle twenty years later.[160]

The building that was once the Lincoln Mill village commissary still stands on Meridian Street. Until recently, an antique store and the Renaissance Theatre were located in the structure, now called the Lincoln Center. Robert

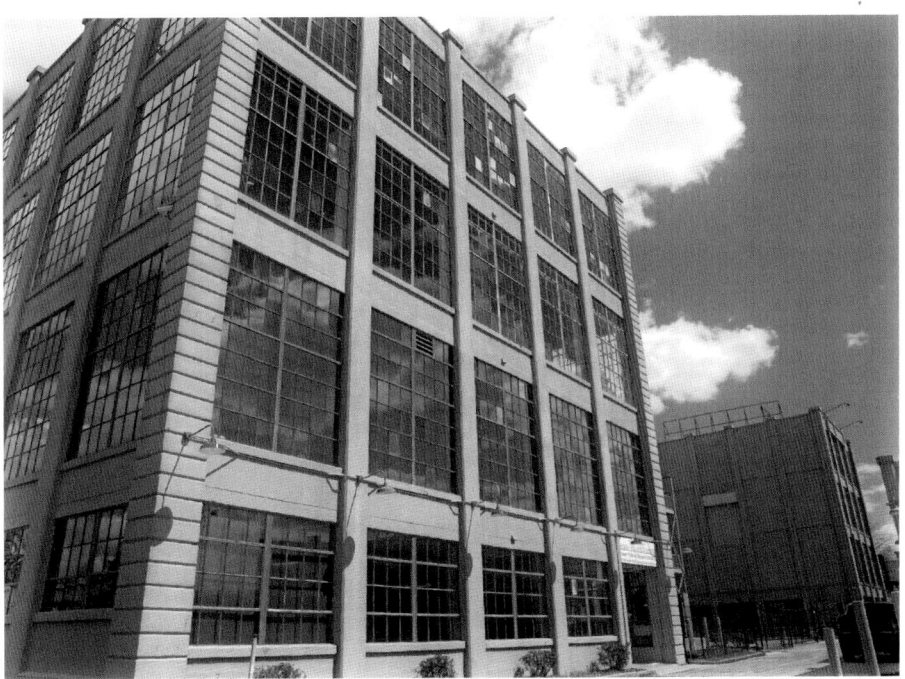

The Lincoln Mill Office Campus today. *Author photo.*

"Bob" Riddle Baker founded the eighty-five-seat theater in 1998 and wrote and produced many of the venue's shows. *Mill Stories*, an original play by Baker, was staged at the theater in 2000. The play, a mixture of historical fact and fiction, was set in Lincoln village in the 1930s. The story's setting was the downstairs café, known in the mill days as the Blue Willow Café. Baker also produced two other mill plays: *Mill Village Christmas* and *Mill Village Memories*.

In 1992, Lincoln Village Ministry started its mission work in the area with a seed, or rather several seeds, planting a one-acre garden to help feed the poor in the impoverished village neighborhood. That tiny seed grew to include education, counseling, medical services, sports programs, housing, Bible study and a church. Lincoln Village Preservation Corporation (LVCP) was founded in 2002 to purchase and renovate homes in the area, providing safe and affordable low-income housing. A team of volunteers from neighboring churches, led by housing director Dale Bowen, works to salvage the neglected and abandoned duplexes into single-family dwellings. Some residents were living in deplorable conditions, according to Bowen. "There was no insulation in any of the houses and no upgrades to electrical or plumbing. They were as they were when they were bought from the mill," he said. Two additional buildings were renovated and provide rental office space for the administrative staff of HEALS Incorporated and LINC Research, a HUBZone (Historically Underutilized Business Zone) business, which must employ 35 percent of its workforce from within a low-income area. A newly renovated community center provides a place where neighborhood residents come for fun, fellowship and learning opportunities. Bowen elaborated, "We provide financial training, education, assistance and support to try to get them to a point where home ownership is a possibility Our primary purpose is to show these people they're valued and made in the image of God, that's why we're here. We do that through education, housing and assistance."

"One of the first things we did was take down the fences. It gives it more of a community feel," said Bowen. Gentrification is not something Bowen or Lincoln Village Ministry wants to see happen in the Lincoln Mill district. Their desire is to better the lives of the people who are already there. "It's kind of like we adopted the area," he said.[161]

The LVCP purchased the Lincoln Elementary School building on Meridian two years after it closed in 2010 to help broaden its scope of community service. Besides housing the Lincoln Village Ministry, the building is also home to Lincoln Academy, a private school that seeks to provide quality Christian education to students zoned to public Title

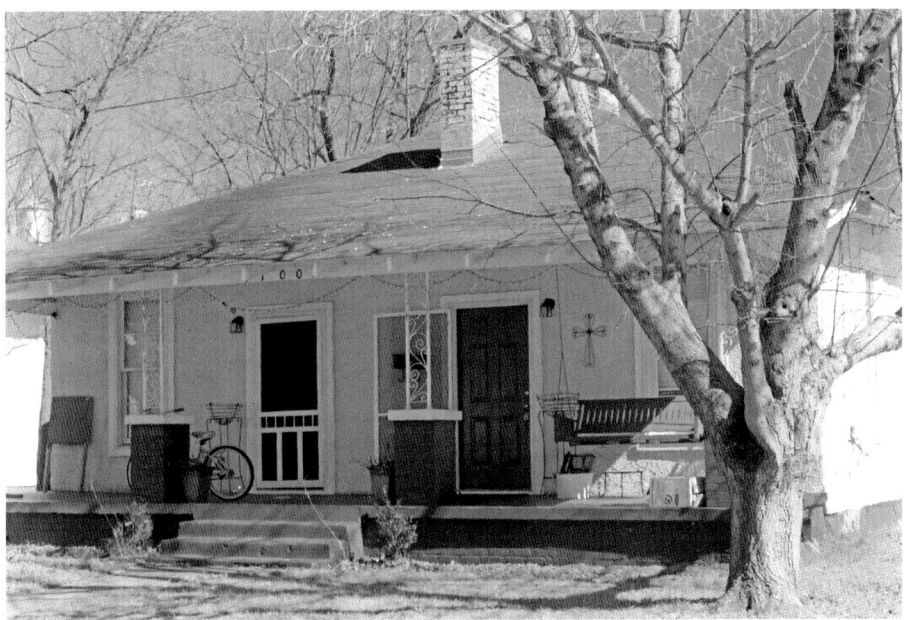

One of the duplexes renovated by Lincoln Village Ministry as a single-family dwelling. *Author photo.*

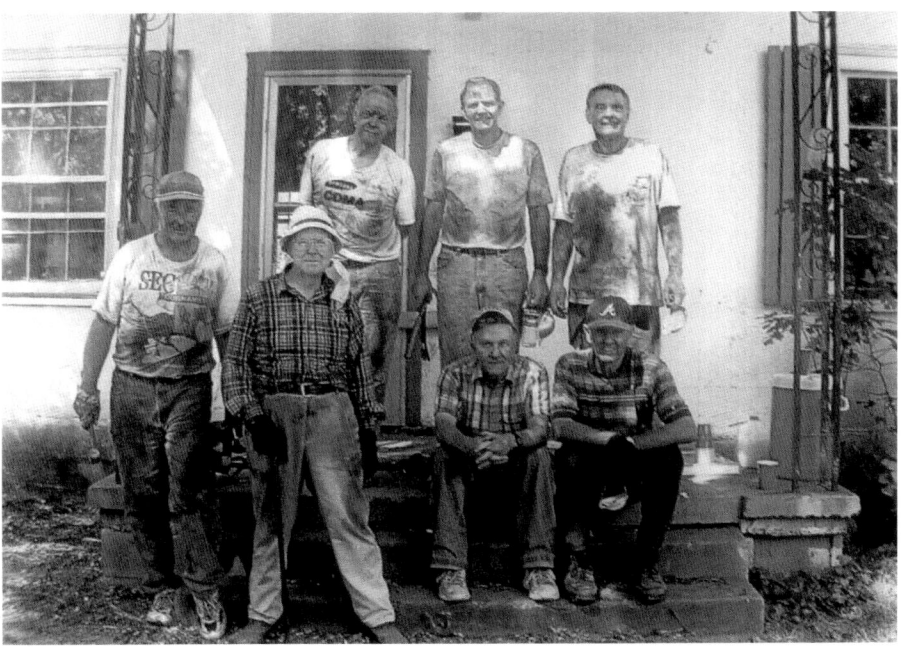

Volunteers with Lincoln Village Ministry take a break after a day of home renovations. *Courtesy of Lincoln Village Ministry.*

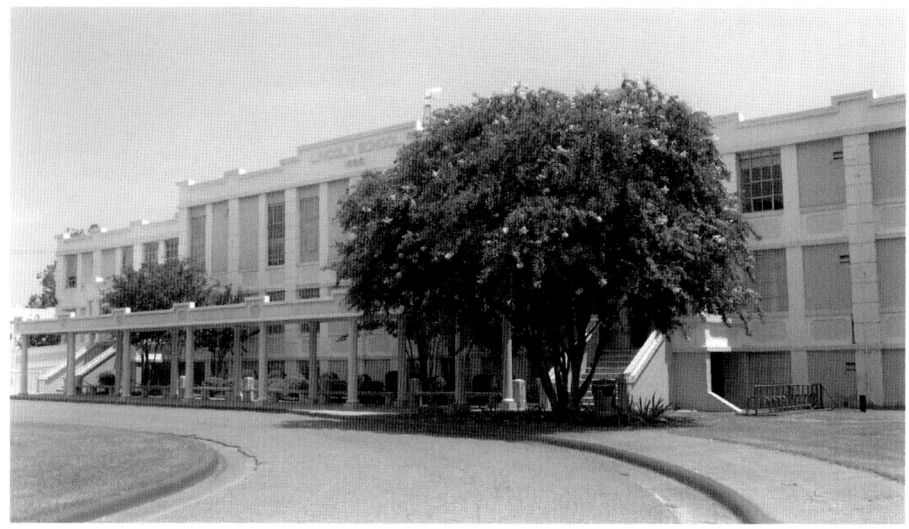

The old Lincoln Elementary School building is now home to Lincoln Academy and Lincoln Village Ministry. *Author photo.*

1 schools, which are those schools that receive federal money to offset neighborhood poverty. Students receive scholarships based on income. The scholarships are granted from RocketSGO and Scholarships for Kids, state scholarship granting organizations. Scholarships may only be granted on behalf of students who meet the eligibility requirements of the Accountability Act.[162]

MERRIMACK MILL DISTRICT

The nearly three hundred structures in the Merrimack Mill village district are bounded, more or less, by Drake Avenue to the south, Dunn Drive to the north, Grote Street to the west and Triana Boulevard to the east. It was declared one of the best old house neighborhoods in the South by *This Old House* in 2012.[163] Unlike years past, the mill villages are no longer homogenized—multicultural families live alongside children and grandchildren of the original owners. Many young couples are finding the diversity and old neighborhood feel in the mill districts appealing.

In 2012, Ryan and Brittney Saffell purchased their home in Merrimack Mill village from a niece of the original owner, the Owens family. The couple spent a year knocking down walls, peeling off layers of wallpaper, sanding and refinishing the original wood flooring and other tasks to make the former duplex a livable one-family dwelling. "It was a hot mess when we moved in," says Brittney. But, they say, the house was structurally sound. "Six by eight rough cut beams on eight-foot piers, I mean they don't make houses like this anymore....Think of all the tornadoes that have happened and these houses still stand here. I think the only thing that brings them down is wood rot and fire." Ryan, an engineer, is now president of the Merrimack Mill Village Organization and says it is important to keep the historical aspect of the community alive. "The homes have to remain recognizable by someone who first lived in them to qualify for historic register status. Ninety percent of the homes in Merrimack Village qualify," he said. "It was important to us to keep the historic integrity intact," said Brittney, who has a degree in interior design. "When people walk in, we still want them to feel like they are in an old house. I think we have managed a balance of old and new well."[164]

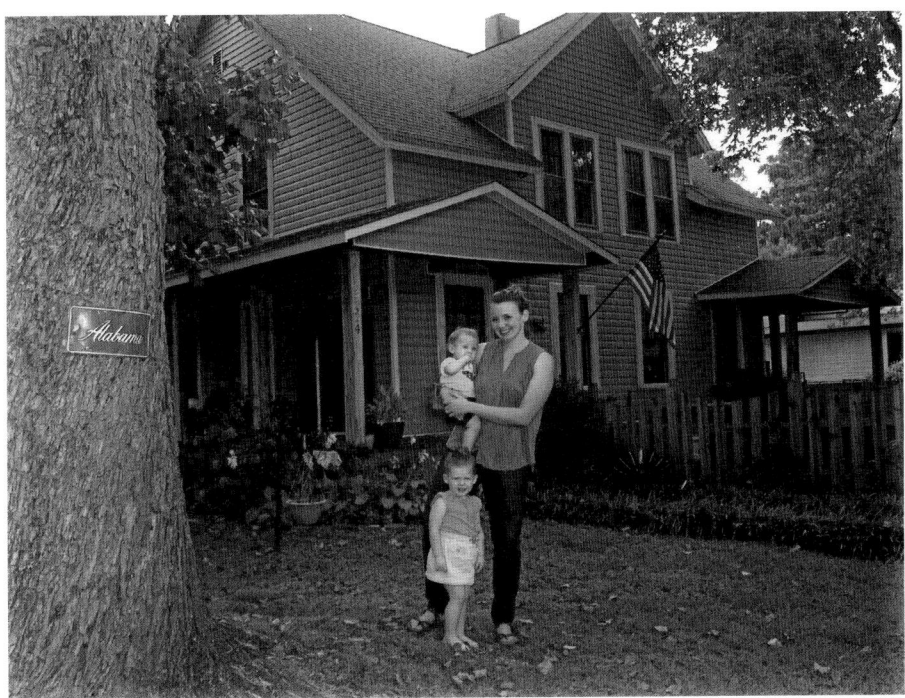

Ryan and Brittney Saffell are raising their children in a century-old mill home in the Merrimack Mill historic district. *Author photo.*

The Saffells and their two small children, along with other families in the village, can enjoy pleasant southern afternoons at Jim Marek Park, a picnic area, playground and pavilion located in the center of the village. The park is named in honor of Merrimack Mill village resident and community leader James J. "Jim" Marek Jr., who passed away in May 2013. Marek founded the Merrimack Mill Village Organization in 2006 to help preserve textile workers' homes in the area. He also worked to restore the Merrimack Mill Village Cemetery with a grant from the Alabama Historical Commission. Under Marek's tenure, street lamps were installed, and he rallied to get a neighborhood watch program and cleanup events started. Marek's campaign efforts were instrumental in getting Merrimack Mill village added to the National Register of Historic Places. The Historic Huntsville Foundation (HHF) established a grant in 2013 to honor the memory of Marek, who served as HHF board chairman. The Jim Marek Grant for Neighborhood Improvement provides $500 grants to two neighborhood organizations, members of the Huntsville Council of Neighborhood Associations, to help develop programs, update playgrounds, improve common areas and give yard beautification awards.[165]

Marek's many achievements on behalf of his community were inspired by the renovation of Merrimack Hall for Performing Arts Center, the old village community center and store, purchased by Alan and Debra Jenkins in 2006. The nonprofit organization, which opened in 2007, provides visual and performing arts education and cultural activities to children and adults with special needs. It was important to the Jenkinses to use the facility not simply as a place for entertainment but also to fulfill a need within the community that was not being met by the schools or service providers. Once the building was a place where lintheads—people not accepted in other areas of the city—could come and purchase goods and enjoy themselves; now it is a haven for another group, not always accepted or appreciated by society, to learn and laugh.

The building housed various tenants after the mill's closing and was a fabric store until the mid-1990s. The Jenkinses did extensive research in order to restore the structure to its original look. Historical photos hang on the walls as well as photos of students and of the many performances that have taken place in the 320-seat theater, which was once the gymnasium. "They used to play indoor baseball in here," said Debra Jenkins. "I can't tell you how many baseballs we found when we took the old sheetrock down." But it seems baseballs weren't all that was uncovered. Staff and volunteers believe that the spirit of child labor poster boy Charlie Foster still visits the old gymnasium.

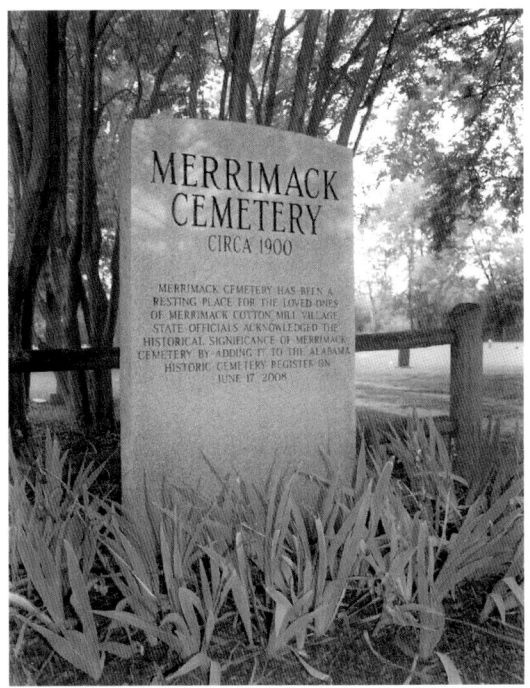

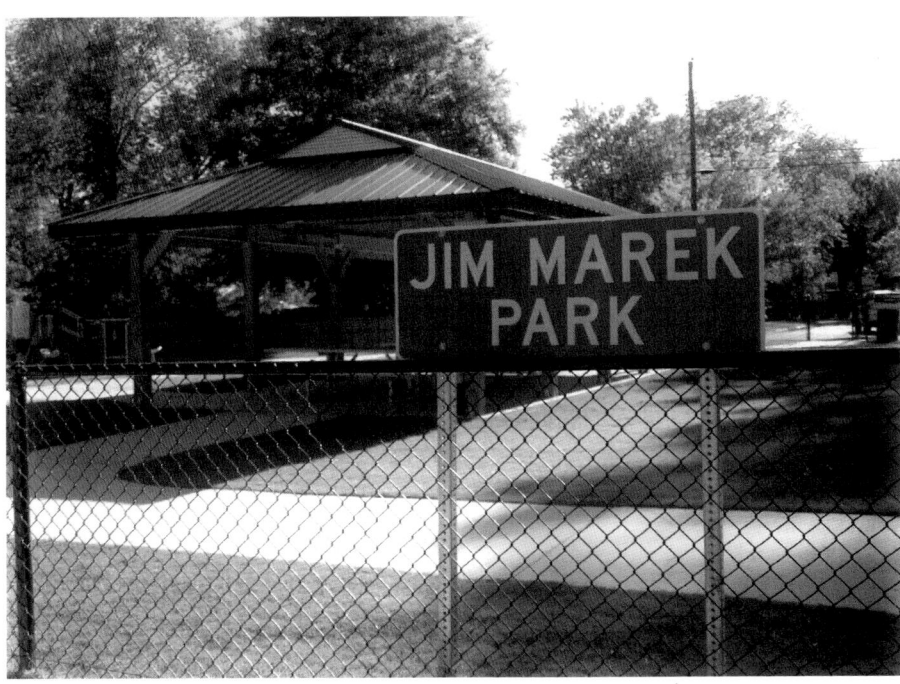

Left: Merrimack Cemetery commemorative marker. *Author photo.*

Below: Jim Marek Park, named in honor of Merrimack Mill village resident and community leader James J. "Jim" Marek Jr. *Author photo.*

"We hear whistling around here all the time," says Jenkins. "Alabama Paranormal Society has spent the night here three times and they describe a presence that lives here that wears overalls and is the protector of everybody. I don't know what I believe, but I like to think that Charlie Foster hangs out here with us." Jenkins believes this was confirmed when she received a call from Foster's niece, who lives in Athens. She had seen her uncle's childhood photo on a poster advertising the production *Upon Their Shoulders: The Merrimack Story* by Huntsville playwright Ron Harris. "She said he wore overalls all of the time, like in the poster. He died in the 1950s of brown lung disease from working in the mill and they buried him in his overalls…and apparently he whistled all of the time."[166]

Charlie Foster and other children of the village once played kick-the-can in the alleys behind their homes; now children can kick a ball on one of ten fields at the Merrimack Soccer Complex on Triana Boulevard near the site of the old Joe Bradley School. While boys used to climb the fence to swim in the lake at the mill's water source, Brahan Spring, all kids can now get wet at the "Everybody Can Play" splash pad aquatic play area. The City of Huntsville and local Kiwanis and Optimist Clubs came together to

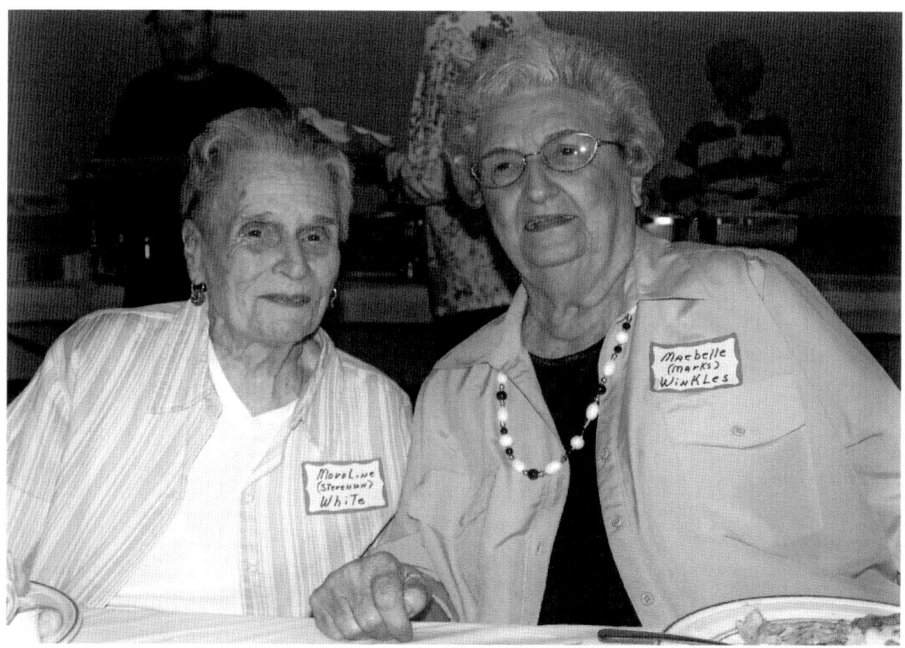

Longtime friends Movoline (Stevenson) White (*left*) and Maebelle (Marks) Winkles enjoy themselves at the Joe Bradley/Merrimack village reunion. *Author photo.*

renovate and improve Brahan Springs Park, located off of Drake Avenue. The playground and water feature are meant to be used by all children, including those with physical limitations. Next to the park, the Brahan Springs Recreation Center was built in 1979. The architectural design of the center is the first and only one of its type in the city. It conserves energy through the use of earth berms around the structure. The center is open to the public, hosts many programs and classes and is home for the Huntsville International League and American League basketball activities. The recently expanded Brahan Spring Natatorium, located at 2213 Drake Avenue, includes two Olympic-sized pools and a therapeutic pool for people with arthritis and other mobility problems. The forty-year-old natatorium includes a modern block and glass façade and a motorized roof that can be rolled back on warm days.[167] The area has come a long way since the days of kick-the-can and illicit swims in the lake.

LOWE MILL DISTRICT

The original hardwood floors of the old Lowe Mill are still stained with oil and marred by indentations made by the massive looms they once supported. Tacks that fell during the making of combat boots are still embedded deep in the wood. Rorschach-like paint splashes and bits of clay are now being added to the strata, forming yet another layer in the history of the building.

When Jim Hudson, founder of Hudson Alpha Research Genetics, visited the Torpedo Factory in Alexandria, Virginia, in the 1970s, his dream to open a similar arts and entertainment center in Huntsville began. Almost thirty years later, that dream came true with the 2001 purchase of Lowe Mill, at the corner of Seminole Drive and Ninth Avenue. The old mill buildings are now home to the largest privately owned arts complex in the United States, Lowe Mill ARTS & Entertainment. Of the complex, Hudson said:

> *The millworkers of the past were hardworking, skilled craftsmen, as are the artists and entrepreneurs of today's Lowe Mill....For over 150 years Huntsville was built around cotton and cotton products. I felt the best way to honor this history of Huntsville's cotton was through preserving the name. Repurposing the structure required many changes. However, preserving the architectural charm of the mill's buildings was very important.*[168]

Huntsville Textile Mills & Villages

These huge historic factory buildings have been redeveloped into 132 working studios for over two hundred artists and makers, including the Flying Monkey Arts Center, fine art galleries, a multi-use theater and performance venues. To operate at Lowe Mill, artists and entrepreneurs must submit an application, which goes through an anonymous jury process. The popular art complex is home to several of north Alabama's premier entertainment events, including Concerts on the Dock and the world's longest-running Cigar Box Guitar Festival. Lowe Mill also highlights the culinary arts, featuring a chocolatier and confectioner, a gourmet popsicle shop, a whiskey distillery, an artisan teahouse, a coffee house, a sandwich shop and a vegetarian food truck. During growing season, local farmers sell produce in the art center's parking lot. Lowe Mill tenants even maintain their own community garden and share the harvest, much like the villagers of earlier years. A bike shop has taken over the old overseer's building. One of the mill's original water towers, still an icon of city progress, is lit with a rainbow of changing colors every evening and can be seen from Memorial Parkway and other areas of the city.

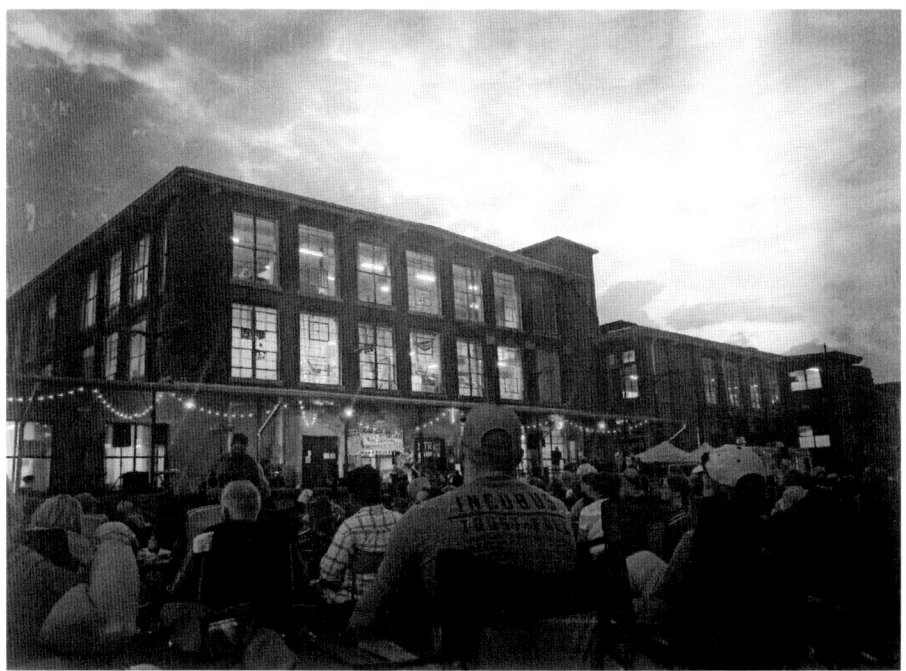

Concerts on the Dock at Lowe Mill ARTS & Entertainment, spring 2016. *Author photo.*

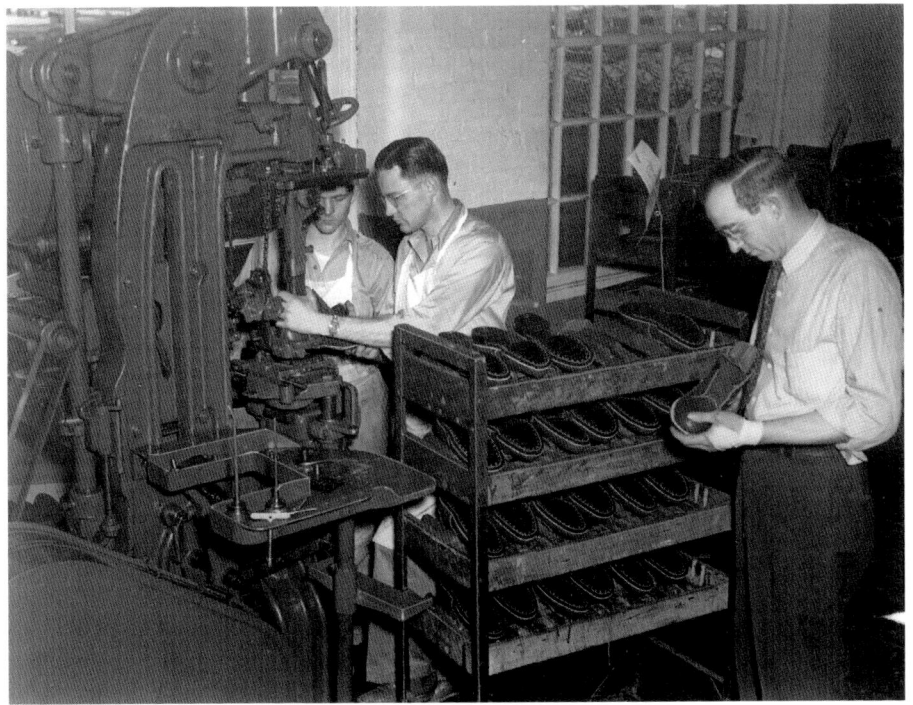

General Shoe Company workers making boots, 1960s. *Huntsville–Madison County Public Library Archives.*

In 2012, Jim Hudson received one of Historic Huntsville Foundation's Pioneer Awards for his restoration and adaptive reuse of Lowe Mill. The most recent expansion of the center was named best downtown project for 2014 by Downtown Huntsville Incorporated. "With our latest expansion we will have repurposed over ninety percent or well over 200,000 square feet of the original structure," said Hudson.

Former employees from the General Shoe/Genesco era come together once a year to reminisce and share stories at a reunion to honor them held in Lowe Mill's first-floor connector space. The event features a 320-square-foot mural painted by Lowe Mill artist Logan Tanner that recounts the history of the mill from a textile factory to a creative workspace. Stanley Keith, a Genesco supervisor in the 1970s, speaks proudly of the boots the company made:

> *The first two and a half years we made every pair of combat boots going over to Vietnam. I presented the millionth pair to General Westmoreland....*

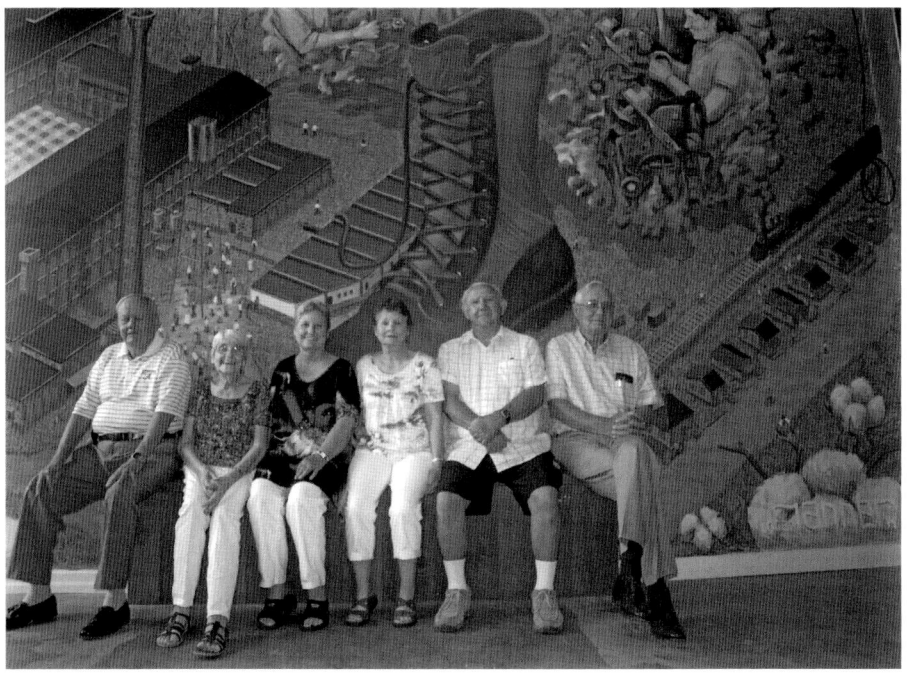

Former Genesco employees (*left to right*) Bill Pendergast, Geraldine Walker, Jean Hinshaw, Mary Johnson, Wayne Johnson and Stanley Keith enjoy getting together once a year for a reunion at Lowe Mill ARTS & Entertainment. Mural by Lowe Mill artist Logan Tanner. *Author photo.*

> *The old boots they had, in a month to six weeks, would tear up in the wet jungles. Our boots would hold up in all the rain, mud and swamps.*

The former Genesco employees have a language all their own, relating stories that took place in the cutting, fitting, lasting and other areas of the plant. Bill Pendergast, whose mother had been a Dallas Mill spinner, started with the Lowe Mill Genesco location in 1957 and stayed with the company after the site closed in 1978, retiring from Genesco in 2004. Mary and Wayne Johnson met at the plant and have been married fifty years. Stanley, Bill and Wayne are the last living former supervisors of the Lowe Mill location. "I hope we are all still here next year," said Keith.[169]

West Huntsville is experiencing a cultural and culinary reawakening. Besides Lowe Mill ARTS & Entertainment, the old school has also been redeveloped into a place where people from all over the city can come and have a good time. Stone Middle School on Clinton Avenue had its beginnings as S.R. Butler High School. The school burned in 1983, was rebuilt in 1986 and closed in 2009.[170] In 2014, developer Randy Schrimsher purchased the property from the city to create a unique development that would preserve the legacy of the school and offer a dynamic entertainment venue for the community. The completely updated building, designed by Matheny Goldmon Architects, is now known as Campus No. 805, an entertainment complex with breweries, restaurants and more. The project got its name from the last three digits of the area's zip code, 35805. The former schoolyard is now a nearly two-acre performance lawn and food truck area where visitors can picnic and listen to music.

More than ever before, the mill districts are becoming vital to the overall spirit and progress of the city of Huntsville. Future plans include a huge pedestrian suspension bridge across Memorial Parkway, part of a bigger project to connect Lowe Mill to the downtown corridor. "This is going to be an art piece that crosses our Parkway….It will be very iconic for the city," said Kathy Martin, City of Huntsville director of engineering.[171]

According to *Alabama Heritage* magazine, Lowe Mill village is one of the state's most outstanding and intact examples of an early twentieth-century mill village:

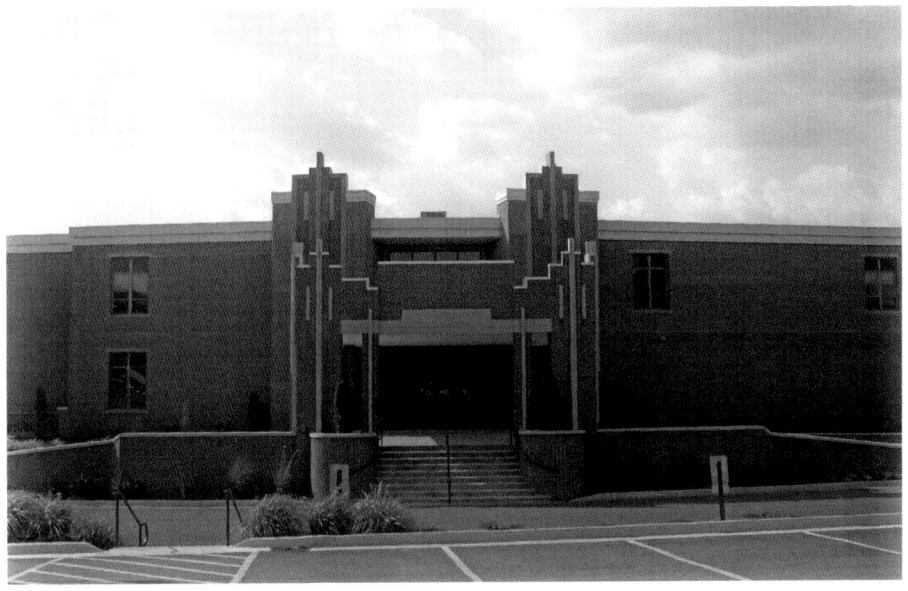

Stone Middle School, which was once S.R. Butler High school, is now home to Campus No. 805, an entertainment complex with breweries, restaurants and more. *Author photo.*

The art deco 1940s Centre Theatre and Drugstore and former J.C. Brown's Mercantile are two of the most significant existing buildings associated with the village. The area has suffered from improper zoning and development, demolitions within the neighborhood, and practice burnings by the Huntsville Fire Department.[172]

Though some of the area has suffered from neglect and misuse, there are individuals and businesses coming in, purchasing houses and renovating them into homes anyone would be proud to live in. Locally owned McNully Properties LLC "rescues and renovates" homes for rental purposes. Rick McNully sold his house and farm in Arkansas to move to Huntsville and join the family-owned business. "I've never done anything like this, but I'm all in," says the master plumber, who's become a jack-of-all-trades. "Back in 1928 I don't think they had tape measures, levels or squares," he says. The McNullys have purchased several houses in the Lowe Mill village. While some landlords, he says, simply "put lipstick" on a house to get it up to code, they completely renovate the structure. "I want to save as many of these old homes as we can.…We are going to try to do something different with each one of them. We want each house to be an individual house," he said. McNully was surprised to find that the attic of one of the homes was insulated with cotton saturated with the soot from the coal-burning fireplace. "It's a wonder it hadn't gone up in flames," he said.

Sometimes there's not much salvageable inside the old homes; still, McNully wants to be sure every one of his rentals holds on to at least a remnant of its historical significance. Original pieces of wood salvaged from the houses were given to Rusted Willow, a woodworking studio at Lowe Mill, to turn into one-of-a-kind art pieces that will return to hang on the walls of the homes.

The McNullys hope to help create a neighborhood where families once again feel safe and where children can ride their bikes and play without fear. "I don't want to see anyone lose their home or be pushed out, but I do want the neighborhood to clean up," he said. Recently they hosted a barbecue cookout at the site of their first renovated home on Seminole Drive. Neighbors along with the west side police force attended the gathering. "The best way to make a difference is by example. It's really that simple," said McNully.[173]

Along with the renovation of some of the older houses, new homes and retail establishments are also in the works for the west side of Huntsville. The city recently sold property across Seminole Drive from the mill to developer

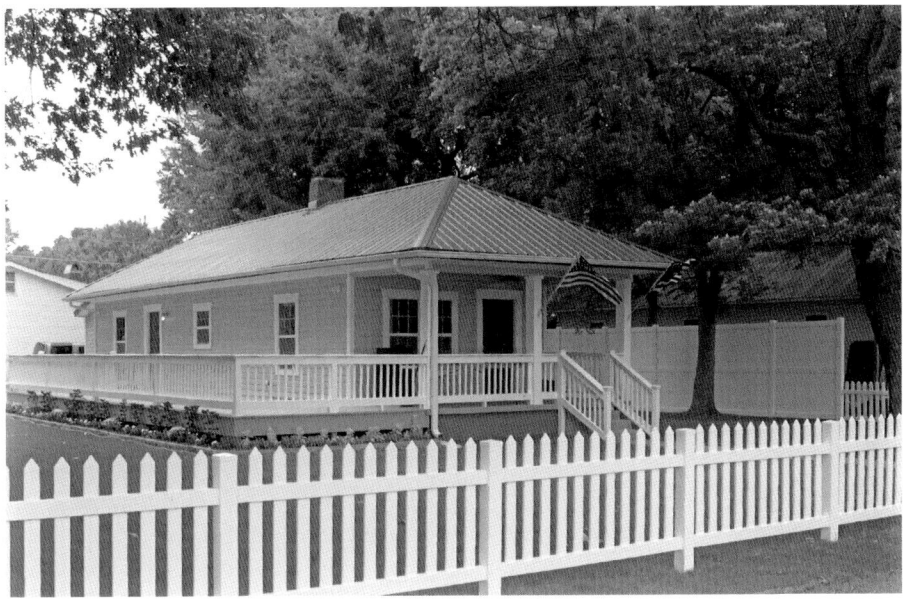

McNully Properties LLC is rescuing and renovating several homes in the Lowe Mill village district. *Author photo.*

Invent Communities in a joint effort to revitalize the area. The company, which will operate in the area as Invent Huntsville LLC, has plans to build residential and mixed-use retail developments. In a recent interview with WHNT Channel 19 news, city administrator John Hamilton had this to say with regard to the revitalization project and the importance of maintaining the historical integrity of the area:

> *We hope that in terms of architecture and the scale of the neighborhood. It looks a lot like it already does. It's part of the history and fabric of Huntsville.... This is not someone coming in and replacing existing homes with a completely different style or size house, this is someone coming in and filling in the gaps.*[174]

Michelle McMullen, president of the West Huntsville Civic Association, was one of the first to meet with the developer. She stressed the importance of getting likeminded people on board to establish and accomplish goals in order to revive the historic village. McMullen has a personal interest in the Lowe Mill district; not only did she grow up in the area, but also her great-grandfather was a local brick mason and made many of the bricks for the mill building. "We have had to reclaim our name and our identity," said McMullen. "I will do everything in my power to save this neighborhood. We need to do it for future generations."[175]

The Legacy of the "Lintheads"

To quote Jessica White, Huntsville Preservation consultant, "We shouldn't just be in the business of saving buildings, we should also focus on saving the communities that helped to build them." These communities were made up not only of mill owners and managers but also of school leaders and coaches, medical staff and everyday working men, women and even children. T.B. Dallas, "Hub" Myhand, Cecil Fain, Phillip Peeler, Edward Anderson, Tracy Pratt, "Big Joe" Bradley, E.F. Dubose, Homer Crim and many other leaders in the textile mill factories, villages and schools made a difference by example. They worked hard to make a living and equally as hard to make a community. That sense of community—of neighbor helping neighbor through both hard times and good times—is the true legacy of the linthead. It's a heritage that is being proudly remembered, honored and revived in these districts today.

CHRONOLOGY

1886	North Alabama Improvement Company (NAIC) forms
1891	T.B. Dallas begins construction of Dallas Mill
1892	Dallas Mill begins operation
1899	Tracy Pratt begins construction of Merrimack Mill
1900–02	Madison Spinning Company opens
1900	Arthur Lowe begins construction of Lowe Mill and West Huntsville Mill
1902	Merrimack opens first company store
1903	Madison Spinning Company closes
1908	Madison Spinning Company reopens as Abingdon Mills
1911	Lowe Mill is purchased by Hunter Manufacturing Company, continues as Lowe Mill
1913	Merrimack opens first hospital and Doutheboys Hall School

Chronology

1915	West Huntsville YMCA opens
1916	West Huntsville's first school is constructed
1918	Abingdon Mills goes bankrupt, is purchased by T.B. Barrell and reopens as Lincoln Mill
1919	Merrimack Mill's Joe Bradley School begins construction
1921	Dallas Mill's Rison School begins construction
1928	Lincoln Mill store/community center and larger school are constructed
1932	Lowe Mill declares bankruptcy, starts up again as Lowe Mills Incorporated
1934	United Textile Workers of America national strike begins in Alabama
1937	Lowe Mill becomes a cotton warehouse
1937–38	Merrimack Mill temporarily closes
1945	Lowe Mill purchased by General Shoe Company/Genesco Incorporated
1945–46	Merrimack Mill sold to M. Lowenstein & Company, reopens as Huntsville Manufacturing Company
1949	Dallas Mill closes
1951	S.R. Butler School opens in West Huntsville to house all four mill school students
1955	Dallas Mill is purchased by General Shoe Company/Genesco Inc.
1955	Lincoln Mill closes

Chronology

1957	Huntsville Industrial Center purchases Lincoln Mill, reopens as Huntsville Industrial Center (HIC)
1978	Martin Industries purchases Lowe Mill to serve as a warehouse
1980	HIC buildings (Lincoln Mill) partially burn
1986	Merrimack's Joe Bradley School is demolished
1987	Dallas Mill is purchased by Gene McLain to serve as a warehouse
1990s	Dallas' Rison School building is demolished to make room for I-565
1991	Dallas Mill is destroyed by fire
1991	Merrimack Mill is demolished
2001	Lowe Mill is purchased by Jim Hudson and renovated into Lowe Mill ARTS & Entertainment
2007	Downtown Industrial Complex is sold to Jim Byrne and renovated into Lincoln Mill Office Complex
2007	Merrimack Hall for Performing Arts opens in former mill community center/store
2010	Lincoln School closes
2012	Lincoln School reopens as Lincoln Academy also housing Lincoln Village Ministry
2015	S.R. Butler High closes

NOTES

Chapter 1

1. Cotton's Journey, "Story of Cotton."
2. Schlingloff, "Cotton Manufacture in Ancient India," 81.
3. Cotton's Journey, "Story of Cotton."
4. Ibid.
5. Gordon, *Business of America*, 17.
6. Nicholas and Guildford, "Samuel Slater and Francis Cabot Lowell."
7. Civil War Trust, "Civil War History: How the Cotton Gin Contributed to the Civil War."
8. Ibid.
9. Arden, "Textile Industry."
10. Surdam, "Traders or Traitors," 302–3.
11. Beckett, "Emancipation and Empire," 1,419.
12. Surdam, "Traders or Traitors," 303.
13. Cohn, *Life and Times of King Cotton*, 121.
14. Snow, "Cotton Mill City," 245.
15. Surdam, "Traders or Traitors," 304.
16. Cohn, *Life and Times of King Cotton*, 124–25.
17. Stephens, *Historic Huntsville*, 64.
18. Garrett, *Atlanta and Environs*, 41–42.
19. Andrews, *Men and the Mills*, 1–4.
20. Snow, "Cotton Mill City," 253.
21. Simpson, *Some Aspects of America's Textile Industry*, 12.

Chapter 2

22. Wilhelm, *History of the Cotton Textile Industry*, 14–23.
23. Snow, "Industry Rising," 36.
24. Miller, "The Fabric of Control," 490.
25. Clark, *History of Manufactures in the United States*, 398.
26. Riverton Intermediate School, "Riverton School History, 1918–2009."
27. Snow, "Cotton Mill City," 247.
28. Stephens, *Historic Huntsville*, 79.
29. Floyd and Mertins, "Dallas Mill," March 14, 1978.
30. *Huntsville Daily Times*, June 28, 1924.
31. *Alabama Republican*, September 7, 1901.

Chapter 3

32. *Huntsville Weekly Mercury*, "Dallas Mill," November 12, 1890.
33. *Huntsville Weekly Democrat*, December 1899.
34. Hanaw, "5 Generations of Life," 17–20.
35. *Huntsville Weekly Mercury*, May 4, 1892.
36. Pruitt, *Eden of the South*, 85.
37. Ibid.
38. Kaylor, "Trevanion Barlow Dallas," 23–25.
39. *Huntsville Times*, "Dallas Employees May Buy Houses," April 15, 1945.
40. Encyclopedia.com, "Genesco Inc."
41. Pruitt, "Dallas and Merrimack Mills," 30–35.
42. *Huntsville Times*, August 1, 1982, D1.
43. Stewart, "Textile Industry in Huntsville and Madison County, Alabama," 6.
44. Roop, "Vintage Photos and Film."
45. Van Osdell, "Cotton Mills, Labor, and the Southern Mind," 60.
46. "Cow Law Upsets Village Residents," in *Old Huntsville: History and Stories of the Tennessee Valley*, 8.
47. Geraldine Walker, interview with author, August 10, 2016.
48. Epting, *Roadside Baseball*, 134.
49. Easterling, "Professor," *Huntsville Times*, August 6, 1985.
50. Easterling, *A Locust Leaves Its Shell*, 61.
51. Curtis, "Mill Village to Main Street."
52. Lambert, "Dallas Mill," *Huntsville Times*, August 1, 1982.
53. Roop, "Vintage Photos and Film."

Chapter 4

54. White, "Huntsville's First Entrepreneur," Huntsville History Collection.
55. King, Schneider and Entzweiler, "Lincoln Mill and Mill Village Historic District." Information provided via e-mail from Nicole J. Woods, administrative assistant, Historic Preservation Division, Alabama Historical Commission, Montgomery, Alabama.
56. U.S. Supreme Court, U.S. Supreme Court Textile Workers v. Lincoln Mills.
57. *Huntsville Times*, "Fruitless Strike Ended," December 28, 1955.
58. Peck, "Teledyne Brown Engineering."
59. Davis, "New Throne for a Cotton King," 6–7.
60. Windsor, "Lincoln Mill Family Reunion."
61. Anderson, *Lincoln Coloring Book*.
62. Easterling, "Special Reunion Is on Its Way," May, 31, 1991.
63. Anderson, *Reflections of Lincoln Village*.
64. Lovvorn, *Day's Gone By*, 38.
65. Ryan, *Northern Dollars for Huntsville Spindles*, 18.
66. Anderson, *Lincoln Coloring Book*.
67. Pruett, "Looking Back," 6.
68. Gathany, "Vintage Base Ball."
69. Archival information used with permission from 48 WAFF-TV. Footage of interview with former news anchor Missy Ming, date unknown. DVD recording courtesy of Larry Lyons.
70. Bayer and Jones, "Lincoln School."
71. *Huntsville Times*, "Lincoln's Retiring Principal to Be Honored by His Students," July 15, 1965.
72. WAFF 48 archives.
73. YouTube, "Huntsville Industrial Center Building Fire."
74. *Huntsville Times*, "A Year Later, the Cause Is Still Unknown," February 18, 1981.

Chapter 5

75. *Huntsville Weekly Mercury*, "Around the Town," May 1899.
76. Fisk and Jenkins, "Merrimack Mill History," 76.
77. Ryan, *Northern Dollars for Huntsville Spindles*, 25.
78. *Huntsville Parker*, Historical Edition, September 1955, 15.
79. Fisk and Jenkins, "Merrimack Mill History," 82.

80. *Huntsville Parker*, Historical Edition, 13.
81. Pruitt, "Dallas and Merrimack Mills," 34.
82. Fisk and Jenkins, "Merrimack Mill History," 79.
83. Deward "Bill" Davis, Doris Davis and Earlene Brown Davis, interview by author, June 27, 2016. Information also from Brown, *Memoirs of William Deward Brown.*
84. *Huntsville Parker*, Historical Edition, 40.
85. King, Schneider and Entzweiler, "Merrimack Mill Village Historic District."
86. Maebelle Winkles, interview by author, June 10, 2016.
87. *Huntsville Mercury*, Centennial Edition, July 23, 1916, 5.
88. Dale, "Fain Recalls School Milestone," July 25, 1992.
89. Easterling, "It's Still Merrimack to Him," August 8, 1993.
90. Marek, "E.F. Dubose: An Educator and Nurseryman."
91. Deward "Bill" Brown, interview by author, June 27, 2016.
92. Fisk and Jenkins, "Merrimack Mill History," 81.
93. *Huntsville Magazine* (Winter 1976–77): 27–28.
94. Pruitt, *Eden of the South*, 236.

Chapter 6

95. *Huntsville Republican*, March 3, 1900; *Huntsville Weekly Mercury*, June 6, 1900.
96. *Huntsville Weekly Mercury*, August 15, 1900.
97. Lowe Mill and Mill Village Historic District, National Register of Historic Places Registration Form. National Park Service, May 5, 2011. Information provided via e-mail from Nicole J. Woods, administrative assistant, Historic Preservation Division, Alabama Historical Commission, Montgomery, Alabama.
98. *Huntsville Times*, Sesquicentennial Issue, September 11–17, 1955.
99. *Huntsville Daily Times*, June 28, 1924.
100. World Heritage Encyclopedia, "Lowe Mill."
101. McMullen, "Lowe Mill."
102. Betty Owens, interview by author, August 15, 2016.
103. Geocaching, "Historic Lowe Mill."
104. Marshall, "At J.C. Brown General Merchandise Building."
105. Gathany, "J.C. Brown General Merchandise."
106. Little, "More Than 250 Turn Out for J.C. Brown Content Sale."

107. Lowe Mill, National Register of Historic Places, 38–40.
108. Ibid., 40.
109. "Introduction: West Huntsville Y.M.C.A. of 1925," *Historic Huntsville Quarterly of Local Architecture and Preservation*, 54.
110. Reinbolt, "Mill Schools of Huntsville," 27–28.
111. Document of Historical Resources. Compiled from the 1975 Centennial History Committee, Part II. Huntsville–Madison County Public Library Heritage Room collection, 5.
112. Ibid.
113. University of Alabama, "University of Alabama in Huntsville."
114. Owens, interview.
115. Roop, "Doors Closing on Butler High."
116. Roop, "Graduate Says Goodbye."

Chapter 7

117. Wasson, "History of East Huntsville Addition," 6.
118. Van West, "Five Points Historic District," National Register of Historic Places Registration Form. National Park Service, January 14, 2012, Section 8: 108.
119. Marshall, "Five Points Shops on Pratt Avenue Are 'Eclectic, Bohemian' Businesses."

Chapter 8

120. Hymer, "Rison-Dallas History."
121. Shabecoff, "Safety Agency to Review Standards on Cotton Dust."
122. McKelway, "Child Labor in Southern Industry," 17.
123. Anderson, "Child Labor Legislation in the South," 77–93.
124. McKelway, "Child Labor in the Southern Cotton Mills," 1–11.
125. Hymer, "Rison-Dallas History."
126. Anderson, "Child Labor Legislation in the South," 85.
127. Cade, "Lewis Hine's Photography and the End of Child Labor in the United States."
128. Leberman, "Hine in Huntsville," 12–13.
129. McKelway, "Child Labor in the South," 163.
130. Ibid., 160.

131. Doris Brown, interview by author, June 27, 2016.
132. McKelway, "Child Labor in Southern Industry," 21.
133. Weeks, *History of Public School Education in Alabama*.
134. Flynt, *Alabama in the Twentieth Century*, 257.
135. Anderson, "Child Labor Legislation in the South," 89.
136. Tabler, "Just Trembling All Over."
137. Salmond, *General Textile Strike of 1934*, 25.
138. Ibid., 26.
139. Irons, *Testing the New Deal*, 101.
140. Sterne, "Comfort Under Control," 22.
141. Ibid., 18.
142. Flynt, *Alabama in the Twentieth Century*, 140.
143. Snow, "Cotton Mill City," 275.
144. Tabler, "Just Trembling All Over."
145. Carney, "Mill Strike," 40.
146. Tabler, "Just Trembling All Over."
147. Salmond, *General Textile Strike of 1934*, 194.
148. Carney, "Mill Strike," 40.
149. Ibid., 41.
150. Salmond, *General Textile Strike of 1934*, 193.
151. Snow, "Cotton Mill City," 276.

Chapter 9

152. Lovington, "High Cotton," 29.
153. Martinson, "Revitalization and Preservation in Alabama's Textile Mill Villages," 76–77.
154. Donna Castellano, executive director, Historic Huntsville Foundation, interview by author, July 1, 2016.
155. Jessica White, historic preservation consultant, City of Huntsville, interview by author, July 1, 2016.
156. Court Heller, interview by author, August 1, 2016.
157. Doyle, "Family's Donation."
158. Luttrell, "Historical Marker Program Moves Forward with Dedication of Two New Markers," 1–3.
159. Frances Akridge, interview by author, August 5, 2016.
160. Jim Byrne, interview by author, March 20, 2016.
161. Dale Bowen, interview by author, July 19, 2016.

162. Lincoln Village Ministry, "Lincoln Academy."
163. Pandolfi et al., "Best Old House Neighborhoods 2012."
164. Ryan and Brittney Saffell, interview by author, June 30, 2016.
165. Historic Huntsville Foundation, "HHF Proposes Grant for Neighborhood Improvement."
166. Debra Jenkins, Merrimack Hall Performing Arts Center, interview by author, July 20, 2016.
167. Huntsville Parks and Recreation, Recreation Guide.
168. Jim Hudson, Lowe Mill ARTS and Entertainment, interview by author, March 23, 2016.
169. Stanley Keith, interview by author, July 30, 2016.
170. "30 Year Anniversary of Stone Middle School Fire."
171. Lough, "Huntsville to Build Sky Bridge."
172. Alabama Heritage, "Lowe Mill Village and Associated Resources, Huntsville, Madison County, c.1900 (Places in Peril 2001)."
173. Rick McNully, interview by author, July 26, 2016.
174. Davis, "City of Huntsville and Developer Partner for Lowe Mill Neighborhood Expansion."
175. Michelle McMullen, interview by author, January 24, 2017.

BIBLIOGRAPHY

Books

Anderson, Arnold Clinton. *Reflections of Lincoln Village: 1944-1958*. Huntsville, AL: self-published, n.d.

Andrews, Mildred Gwin. *The Men and the Mills: A History of the Southern Textile Industry*. Macon, GA: Mercer University Press, 1987.

Betts, Edward Chambers. *Early History of Huntsville, Alabama: 1804 to 1870*. Montgomery, AL: Brown Printing, 1916.

Broadus, Mitchel. *The Rise of Cotton Mills in the South*. New York: Johns Hopkins Press, 1968.

Brown, William D. *Memoirs of William Deward Brown: Merrimack, the Merchant Marine and Me*. N.p.: self-published, 2005.

Clark, Victor S. *History of Manufactures in the United States: 1860–1893*. Vol. 2. New York: Carnegie Institution of Washington, 1929.

Cohn, David L. *The Life and Times of King Cotton*. New York: Oxford University Press, 1956.

Conway, Mimi. *Rise Gonna Rise: A Portrait of Southern Textile Workers*. New York: Doubleday, 1979.

Daniel, Clete. *Culture of Misfortune: An Interpretive History of Textile Unionism in the United States*. Ithaca, NY: Cornell University Press, 2001.

Easterling, Bill. *A Locust Leaves Its Shell*. Winter Haven, FL: Kaylor & Kaylor, 2000.

Bibliography

Epting, Chris. *Roadside Baseball: The Locations of America's Baseball Landmarks.* Santa Monica, CA: Santa Monica Press, 2009.

Flynt, Wayne. *Alabama in the Twentieth Century.* Tuscaloosa: University of Alabama Press, 2004.

———. *Poor but Proud.* Tuscaloosa: University of Alabama Press, 1989.

Garrett, Franklin M. *Atlanta and Environs: A Chronicle of Its People and Events, 1880–1930s.* Vol. 2. Athens: University of Georgia Press, 1969.

Gordon, John Steele. *The Business of America: Tales from the Marketplace—American Enterprise from the Settling of New England to the Breakup of AT&T.* New York: Walker Publishing Company, 2001.

Hall, Jacquelyn Dowd, et al. *Like a Family: The Making of a Southern Cotton Mill World.* Chapel Hill: University of North Carolina Press, 1987.

Huntsville Alabama Sesquicentennial Commemorative Album 1805–1955. Huntsville, AL: 1955.

Irons, Janet. *Testing the New Deal: The General Strike of 1934 in the American South.* Urbana: University of Illinois Press, 2002.

Lovvorn, J. Curtis. *Day's* [sic] *Gone By: Cotton Mill Village Life.* Woodville, AL: Home Town Publishers, 2008.

Pruitt, Renee G. *Eden of the South: A Chronology of Huntsville, Alabama, 1805–2005.* Huntsville, AL: Huntsville–Madison County Public Library, 2005.

Salmond, John A. *The General Textile Strike of 1934: From Maine to Alabama.* Columbia: University of Missouri Press, 2002.

Simpson, William Hays. *Some Aspects of America's Textile Industry: With Special Reference to Cotton.* Columbia: Division of General Studies, University of South Carolina, 1966.

Stephens, Elise Hopkins. *Historic Huntsville: A City of New Beginnings.* Sun Valley, CA: American Historical Press, 2002.

Wiggins, Sarah Woolfolk, ed. *From Civil War to Civil Rights, Alabama, 1860–1960: An Anthology from the Alabama Review.* Tuscaloosa: University of Alabama Press, 1987.

Wilhelm, Dwight M. *History of the Cotton Textile Industry of Alabama: 1809–1950.* Montgomery, AL: Dwight M. Wilhelm, 1950.

Journals

Anderson, Neal L. "Child Labor Legislation in the South." *Annals of the American Academy of Political and Social Science* 25 (1905): 77–93.

Bibliography

Beckett, Sven. "Emancipation and Empire: Reconstructing the Worldview Web of Cotton Production in the Age of the American Civil War." *American Historical Review* 109, no. 5 (2004): 1405–38.

"The Cotton Mills in the Piedmont—Social Conditions of the Operative Improved." *Republican* 2, no. 24 (March 9, 1901).

Fisk, Sarah Huff, and Debra Jenkins. "Merrimack Mill History." *Huntsville Historical Review* 33, no. 1 (Winter–Spring 2008): 75–84.

Kaylor, Mike. "Trevanion Barlow Dallas: His Huntsville Connections." *Huntsville Historical Review* 14, nos. 1–2 (1984): 13–28.

Koistinen, David. "The Causes of Deindustrialization: The Migration of the Cotton Textile Industry from New England to the South." *Enterprise & Society* 3, no. 3 (2002): 482–520.

Leberman, Susannah B. "Hine in Huntsville: What the Photographic Detective Found." *Huntsville Historical Review* 29, no. 1 (Fall–Winter 2003).

Luttrell, Alex F., III. "Historical Marker Program Moves Forward with Dedication of Two New Markers." *Huntsville Historical Review* 22, no. 2 (1995): 1–3.

McKelway, A.J. "Child Labor in Southern Industry." *Annals of the American Academy of Political and Social Science* 25 (1905): 16–22.

———. "Child Labor in the South." *Annals of the American Academy of Political and Social Science* 35, no. 1 (1910): 156–64.

———. "Child Labor in the Southern Cotton Mills." *Annals of the American Academy of Political and Social Science* 27 (1906): 1–11.

Miller, Randall M. "The Fabric of Control: Slavery in Antebellum Southern Textile Mills." *Business History Review* 55, no. 4 (1981): 471–90.

Pruett, John. "Looking Back: A Sports History of Huntsville." *Huntsville Historical Review* 9 (January–April 1979): 3–25.

Pruitt, Ranee G. "The Dallas and Merrimack Mills." *Huntsville Historical Review* 29, no. 1 (Fall–Winter 2003): 30–35.

Reinbolt, Aida. "The Mill Schools of Huntsville." *Historic Huntsville Quarterly of Local Architecture and Preservation* 12, nos. 3 and 4 (Spring–Summer 1986): 25–31.

Rohr, Nancy. "The O'Shaughnessy Legacy in Huntsville." *Huntsville Historical Review* 21, no. 2 (1994): 1–17.

Ryan, Patricia H. "Tracy Pratt." *Huntsville Historical Review* 15, nos. 1 and 2 (1985): 27–37.

Schlingloff, D. "Cotton Manufacture in Ancient India." *Journal of Economic and Social History* 17, no. 1 (1974): 81–90.

Snow, Whitney Adrienne. "Cotton Mill City: The Huntsville Textile Industry, 1880–1989." *Alabama Review* 63, no. 4 (October 1, 2010): 243–81.

Bibliography

———. "Industry Rising: Madison County Cotton Mills, 1809–1885." *Huntsville Historical Review* 39, no. 2 (2014): 23–38.
Surdam, David G. "Traders or Traitors: Northern Cotton Trading During the Civil War." *Business and Economic History* 28, no. 2 (1999): 301–12.
Terrill, Tom E., Edmond Ewing and Pamela White. "Eager Hands: Labor for Southern Textiles, 1850–1860." *Journal of Economic History* 36, no. 1 (1976): 655–80.
Wasson, Joberta. "History of East Huntsville Addition." *Historic Huntsville Quarterly of Local Architecture and Preservation* (Fall–Winter 1983–84): 3–10.

Magazine and Newspaper Articles

Carney, Tom. "The Mill Strike." *Old Huntsville: History and Stories of the Tennessee Valley*, no. 278, April 2016.
Dale, James. "Fain Recalls School Milestone." *Huntsville Times*, July 25, 1992.
Davis, Louis. "New Throne for a Cotton King." *Nashville Tennessean Magazine*, January 19, 1964.
Easterling, Bill. "It's Still Merrimack to Him." *Huntsville Times*, August 8, 1993.
———. "The Professor." *Huntsville Times*, August 6, 1985.
———. "A Special Reunion Is on Its Way." *Huntsville Times*, May, 31, 1991.
Lambert, Lane. "Dallas Mill: Landmark Has Ghosts, Goats, Place in History." *Huntsville Times*, August 1, 1982.
Leigh, Phil. "Trading with the Enemy." *New York Times*, October 28, 2012.
Little, Jim. "More Than 250 Turn Out for J.C. Brown Content Sale." *Huntsville Times*, June 4, 2015.
Lovington, Sara Wright. "High Cotton." *NO'ALA Huntsville* 3, no. 3 (May–June 2014): 28–37.
Marshall, Mike. "At J.C. Brown General Merchandise Building, It's Like Stepping into the Past." *Huntsville Times*, February 26, 2012.
———. "Five Points Shops on Pratt Avenue Are 'Eclectic, Bohemian' Businesses." *Huntsville Times*, January 21, 2011.
Old Huntsville: History and Stories of the Tennessee Valley, nos. 69 and 92, n.d.
Roop, Lee. "Doors Closing on Butler High after 64 Years in Huntsville." *Huntsville Times*, May 20, 2015.
———. "A Graduate Says Goodbye to Huntsville's Butler High School." *Huntsville Times*, May 22, 2015.
Shabecoff, Philip. "Safety Agency to Review Standards on Cotton Dust." *New York Times*, March 28, 1981.

Bibliography

Sterne, Pamela King. "Comfort Under Control: Alabama's Textile Mill Villages." *Alabama Heritage*, no. 73 (Summer 2014): 15–23.

Windsor, Shawn. "Lincoln Mill Family Reunion—Old Friends Gather for Annual Affair." *Huntsville Times*, June 27, 1999.

Websites and Blogs

Alabama Heritage. "Lowe Mill Village and Associated Resources, Huntsville, Madison County, c.1900 (Places in Peril 2001)." October 21, 2002, accessed August 13, 2016, http://www.alabamaheritage.com.

Arden, William. "Textile Industry." New Georgia Encyclopedia. http://www.georgiaencyclopedia.org/articles/business-economy/textile-industry.

Arkwright Society. "History." *Sir Richard Arkwright's Cromford Mills.* https://www.cromfordmills.org.uk/history.

Bayer, Linda, and Harvie Jones. "Lincoln School." National Register of Historic Places Inventory-Nomination Form. National Park Service. July 2, 1982. http://focus.nps.gov/pdfhost/docs/NRHP/Text/82001608.pdf.

Brooks, Chris. "Ancient Cotton." Center for World History, 2008. http://cwh.ucsc.edu/brooks/The_Ancient_World.html.

Cade, D.L. "Lewis Hine's Photography and the End of Child Labor in the United States." September 7, 2013. http://petapixel.com/2013/09/07/lewis-hines-photography-end-child-labor-united-states.

Civil War Trust. "Civil War History: How the Cotton Gin Contributed to the Civil War." http://www.civilwar.org/resources/civil-war-history-how-the.html.

Cotton's Journey "Story of Cotton." http://www.cottonsjourney.com/storyofcotton/page2.asp.

Davis, Chris. "City of Huntsville and Developer Partner for Lowe Mill Neighborhood Expansion." WHNT News 19, November 19, 2016. http://whnt.com.

Down, Matthew L. "Great Depression in Alabama." *Encyclopedia of Alabama.* http://www.encyclopediaofalabama.org/article/h-3608.

Doyle, Steve. "Family's Donation Will Maintain Goldsmith-Schiffman Field's 'Treasure' Place as Huntsville Sports Venue." al.com, September 5, 2014. http://www.al.com/news/huntsville/index.ssf/2014/09/familys_donation_will_maintain.html.

Encyclopedia.com. "Cotton Kingdom." http://www.encyclopedia.com/doc/1G2-2536600938.html.

Bibliography

Floyd, W. Warner, and Ellen Mertins. "Dallas Mill." National Register of Historic Places Inventory-Nomination Form. National Park Service. March 14, 1978. http://focus.nps.gov/pdfhost/docs/NRHP/Text/78000496.pdf.

Gathany, Bob. "J.C. Brown General Merchandise Founded 1898—Contents to Be Sold Starting Thursday." *Huntsville Times*, June 3, 2015. http://www.al.com/living/index.ssf/2015/06/jc_brown_general_merchandise_f.html.

———. "Vintage Base Ball Comes to Huntsville at Lowe Mill Starting Saturday." al.com, March 17, 2016. http://www.al.com/living/index.ssf/2016/03/vintage_base_ball_comes_to_hun.html.

Geocaching.com. "Historic Lowe Mill." Accessed June 2, 2016, https://www.geocaching.com/geocache/GC19MD4_the-historic-lowe-mill.

Hanaw, Margaret Anne Goldsmith. "5 Generations of Life: 'My Family and the Huntsville Alabama Jewish Community,' 1952–1982." *Huntsville Historical Review* 12, nos. 3 and 4 (July–October 1982): 5–43. http://huntsvillehistorycollection.org.

Historic Huntsville Foundation. "HHF Proposes Grant for Neighborhood Improvement." June 14, 2013, accessed August 9, 2016, http://www.historichuntsville.org/home.

Huntsville Parks and Recreation. Recreation Guide. http://www.huntsvilleal.gov/recreation/HsvRecGuide.pdf.

Hymer, Sarah Ann. "Rison-Dallas History." http://www.rison-dallas.com/history3.html.

"Introduction: West Huntsville Y.M.C.A. of 1925." *Historic Huntsville Quarterly of Local Architecture and Preservation* 15, no. 4 (Summer 1989): 3–56. http://huntsvillehistorycollection.org.

Lincoln Village Ministry. "Lincoln Academy." Accessed August 1, 2016, https://lincolnvillageministry.com/our-ministry/education/lincoln-academy.

Lough, Nick. "Huntsville to Build Sky Bridge over Parkway, Connect Downtown and Lowe Mill." *48 WAFF*, June 7, 2016. http://www.waff.com.

Marek, Jim. "E.F. Dubose: An Educator and Nurseryman." Reprint from the *Merrimack Mill Village Newsletter*, June–July 2011. http://huntsvillehistorycollection.org/hh/index.php?.

Maulsby, Ann Geiger. *Merrimack Cemetery, Huntsville, Alabama*. http://huntsvillehistorycollection.org/hh/hhpics/pdf/book/Merrimack_Cemetery-original.pdf.

McCarter, Mark. "Rebuilding West Huntsville's Mill Villages: Redevelopment Seeks to Preserve and Respect West Huntsville's Rich History and Diversity." HUNTSVILLE *City Blog*, November 27, 2016. https://cityblog.huntsvilleal.gov/rebuilding-lowe-mill-neighborhood.

Bibliography

Nicholas, Tom, and Matthew Guildford. "Samuel Slater and Francis Cabot Lowell: The Factory System in U.S. Cotton Manufacturing." *Harvard Business Review Case Study.* February 2, 2014. http://www.hbs.edu/faculty/Pages/item.aspx?num=46048.

Pandolfi, Keith, Amanda Shettleton, Gillian Barth, Ambrose Martos, Elsa Saatela and Meredith Richards. "Best Old House Neighborhoods 2012: The South." *This Old House.* http://www.thisoldhouse.com/toh/photos/0,,20569037_21121271,00.html.

Peck, Emily. "Teledyne Brown Engineering." *Encyclopedia of Alabama.* May 18, 2009, updated April 24, 2013. http://www.encyclopediaofalabama.org/article/h-2155.

Riverton Intermediate School, Madison County Schools. "Riverton School History, 1918–2009." https://www.madison.k12.al.us/Schools/ris/Pages/AboutUs.aspx.

Roop, Lee. "Vintage Photos and Film." al.com, July 24, 2015.

Stamps, Katie. "Straight to Ale Brewery Integrates Old and New at Former Stone Middle School." *Huntsville Historic Preservation Commission*, September 30, 2015. https://huntsvillehpc.wordpress.com.

Tabler, Dave. "Just Trembling All Over—The Textile Strike of 1934 in Huntsville, AL." *Appalachian History*, August 20, 2004. http://www.appalachianhistory.net/2014/08/just-trembling-textile-strike-1934-huntsville-al.html.

University of Alabama. "University of Alabama in Huntsville." Accessed September 1, 2016, http://www.uah.edu/about/history.

U.S. Supreme Court. U.S. Supreme Court Textile Workers v. Lincoln Mills, 353 U.S. 448 (1957) Textile Workers Union of America v. Lincoln Mills of Alabama, No. 211. https://supreme.justia.com/cases/federal/us/353/448.

Weeks, Stephen B. *History of Public School Education in Alabama.* Bulletin 1915, no. 12. Whole Number 637, 1-212. United States Bureau of Education, Department of the Interior, retrieved from ERIC database. Accessed June 14, 2016. http://eric.ed.gov/?id=ED541810.

White, Gilbert. "Huntsville's First Entrepreneur: The 'Salt King' of Abingdon, VA." Huntsville History Collection. http://huntsvillehistorycollection.org.

World Public Library. "Lowe Mill." *World Heritage Encyclopedia.* http://publiclibrary.net/articles/Lowe_Mill.

YouTube. "Huntsville Industrial Center Building Fire." Uploaded February 3, 2011, accessed June 9, 2016, https://www.youtube.com/watch?v=3XkvbsghgcA.

Bibliography

Miscellaneous

Anderson, Arni. *Lincoln Coloring Book—Historic Village 1900–Present*. Huntsville, AL: Marvin Lovvorn, 2013.

Crabtree, Butch. *Merrimack: A Short History*. Huntsville, AL: Huntsville Public Library, n.d.

A Document of Historical Resources. Compiled from the 1975 Centennial History Committee, Part II. Huntsville–Madison County Public Library Heritage Room collection.

Hodgson, Joseph, ed. *The Alabama Manual and Statistical Record for 1869*. Montgomery, AL: Mail Building, 1869.

King, Pamela Sterne, David B. Schneider and Susan Entzweiler. "Lincoln Mill and Mill Village Historic District." National Register of Historic Places Registration Form. National Park Service. March 8, 2010.

———. "Merrimack Mill Village Historic District." National Register of Historic Places Registration Form. National Park Service. March 1, 2010.

Lowe Mill and Mill Village Historic District. National Register of Historic Places Registration Form. National Park Service. May 5, 2011.

Martinson, Lauren Burlison. "Revitalization and Preservation in Alabama's Textile Mill Villages." Master's thesis, University of Georgia, 1998.

McMullen, Michelle. "Lowe Mill: Huntsville's Forgotten Mill Village." *Foundation Forum: Historic Huntsville Foundation Newsletter* (Winter 2006), 6.

Miles, Diana Greer. "Revitalization in Two Textile Mill Villages in South Carolina." Master's thesis, University of Georgia, 1996.

"Mill Village to Main Street," VHS video, produced and directed by Dick Curtis; videotaped by Blake Hudson; research by Sarah Shouse and Elise Stephens, OCLC Number: 700971682, Huntsville Madison County Public Library.

Ryan, Patricia H. *Northern Dollars for Huntsville Spindles*. Huntsville, AL: Planning Department Special Report No. 4, 1983.

Stewart, Robert. "The Textile Industry in Huntsville and Madison County, Alabama." HAER Southern Textile Industry Survey for Historic American Engineering Record. National Park Service, U.S. Department of the Interior, October 1998.

Van Osdell, John G., Jr. "Cotton Mills, Labor, and the Southern Mind: 1880–1930." PhD diss., Tulane University, 1966.

Van West, Carroll. "Five Points Historic District," National Register of Historic Places Registration Form. National Park Service, January 14, 2012, Section 8: 108.

INDEX

A

Alabama Historical Commission 98
Anderson, Edward 55, 56, 79, 118
Arkwright, Richard 13, 14, 15

B

Barrell, William Lincoln 48, 49
Big Spring 37
Bonner, Wayne 102
Bowen, Dale 104
Bradley, Joseph (Big Joe) 61, 66, 68, 89
Brahan Spring 110
Brahan Spring Natatorium 111
Byrne, Jim 102, 103

C

Cabaniss, Charles 27, 28
Campus No. 805 115
Castellano, Donna 98
Clark, Elijah 38
Coons, Joshua 31, 33, 34
Crim, J. Homer 79, 81, 118
crop-lien system 22, 23

Cummings, Milton 52

D

Dallas, Trevanion Barlow (T.B.) 36, 38, 39, 118
Dean, John 91, 92, 93
Donegan, James J. 28
Dubose, Edward Foyle (E.F.) 68, 69, 71, 79, 118
Dunn, Woodrow E. 62

E

Ebaugh, Robin 57
Echols, W.H. 28
1881 Textile Exposition, Atlanta 23, 24

F

Fain, Cecil V. 43, 44, 45, 68, 69, 79, 118
Foster, Charlie 87, 88

Index

G

Genesco/General Shoe Company 39, 75, 113, 114
Goldsmith, Oscar 40
Graniteville Mill 17
Great Depression 25, 49, 61, 75, 89, 95
Gregg, William 17, 18
Grote, Dr. C.A. 67

H

Hargreaves, James 13
Haughton, William 28
Hine, Lewis 26, 86, 87, 88
Historic Huntsville Foundation 98, 108, 113
Historic Huntsville Preservation Commission 98
Hudson, Jim 76, 111, 113
Huntsville Industrial Center 52
Huntsville Land Company 40
Huntsville–Madison County Historical Society 98
Huntsville Manufacturing Company 62, 63, 64, 71
Huntsville Railway Company 54

J

J.C. Brown General Merchandise 76, 77, 78
Jenkins, Alan 108
Jenkins, Debra 96, 108, 110
Jones, Horatio 27, 28

L

Lincoln Academy 56, 104, 106
Lincoln Mill Ministry 104, 105
Lincoln Village Preservation Corporation 104
lintheads 26, 41, 54, 108, 118
Lowe, Arthur H. 73, 74
Lowell, Francis Cabot 14, 16
Lowell, Massachusetts 14, 16, 25, 35, 59
Lowe Mill ARTS & Entertainment 111, 114
Lowenstein & Company 62, 71, 72

M

Marek, Jim 108, 109
McCormick, Mary (Virginia) 77, 79
McLain, Gene 39, 47, 76
Merrimack Hall for Performing Arts 108
Merrimack Soccer Complex 110
Myhand, H.E. (Hub) 41, 42

N

National Register of Historic Places 97, 98, 99, 108
National Rights Administration (NRA) 90, 91
North Alabama Improvement Company (NAIC) 33, 36, 82
Northwestern Land Association (NLA) 36, 82, 83

O

O'Shaughnessy, James 31, 32, 33, 36, 38
O'Shaughnessy, Michael 32, 33, 36, 38

P

Patton, Donegan and Company 28, 29
Patton, William 28, 30
Peeler, Phillip W. 49, 50, 55, 56, 118
Poor, Charles 74
Pratt, Tracy W. 31, 33, 34, 35, 36, 39, 59, 60, 64, 73, 82, 83, 118

R

Renaissance Theatre 103
Rison, Archie L. 38, 43
Rison, William R. 38, 87

INDEX

S

Slater Mill 15
Slater, Samuel 14, 15, 16

U

United Textile Workers of America
 (UTW) 91, 92

V

von Braun, Wernher 96, 103

W

Ward, James A. 36, 82
Wellman, Willard I. 31, 33, 35, 36,
 39, 73, 82, 83
Wells, William S. 31, 33, 36, 39, 82,
 83
White, James 48
White, Jessica 98, 99, 118
Whitney, Eli 16, 17

ABOUT THE AUTHOR

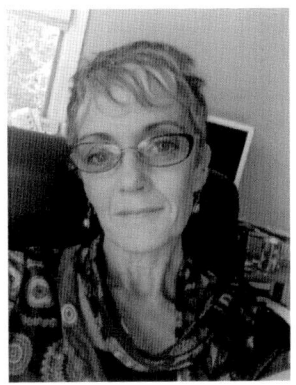

Terri L. French is an award-winning poet and writer with an interest in regional culture and history. She has worked as a licensed massage therapist, a freelance writer and an editor. Terri has also served as southeast regional coordinator for the Haiku Society of America and secretary for the Alabama Writers Conclave, and she is currently the secretary for the Haiku Foundation. The native Michigander has lived in the Tennessee Valley since 1987. She and her husband, Ray, a NASA engineer, have four mostly grown children, a spoiled dog and two rotten cats.

Visit us at
www.historypress.net

This title is also available as an e-book